LABORATORY
EXPERIMENTS
IN CHEMISTRY

PUBLISHER'S
NOTE

Laboratory Experiments in Chemistry is suitable for use in conjunction with *Chemistry: A Conceptual Approach, Third Edition* or *Introduction to Chemistry*, both by Professor Charles E. Mortimer.

The Instructor's Manual that accompanies this laboratory manual suggests alternative sequences in which *Laboratory Experiments in Chemistry* may be used with either one of Professor Mortimer's textbooks.

LABORATORY EXPERIMENTS IN CHEMISTRY

JOSEPH J. LAGOWSKI
STEPHEN E. WEBBER

University of Texas, Austin

D. VAN NOSTRAND COMPANY
New York Cincinnati Toronto London Melbourne

D. Van Nostrand Company Regional Offices:
New York   Cincinnati

D. Van Nostrand Company International Offices:
London   Toronto   Melbourne

Copyright © 1977 by Litton Educational Publishing, Inc.

Library of Congress Catalog Card Number: 76-62887
ISBN: 0-442-24653-6

Published by D. Van Nostrand Company
450 West 33rd Street, New York, N.Y. 10001

10 9 8 7 6 5 4 3 2 1

PREFACE

This laboratory manual is the result of our combined experience in coordinating a large laboratory course for freshmen with a wide diversity of backgrounds. Our objectives in *Laboratory Experiments in Chemistry* are:

(1) to train students in elementary techniques of chemistry laboratory work, thereby providing the basis for more advanced work;

(2) to train students to organize their laboratory work efficiently and to work accurately in the laboratory (with good understanding of the sources of potential experimental error);

(3) to deepen the student's understanding of chemical phenomena by direct experience.

The experiments are divided into two groups — *Introductory Experiments* and *General Experiments*. We recommend that all beginning students do the introductory set to learn the basic skills required for later experiments. The format for tabulation of data and other observations are explicit in the introductory section in order to provide a model for proper record keeping in laboratory work.

The General Experiments are grouped according to type. They may, of course, be performed in any order compatible with the course sequence. The more advanced experiments are indicated (in the contents) by an asterisk. We believe that thirty-seven experiments in this manual give the instructor sufficient flexibility to vary the course content from semester to semester.

Each experiment is organized as follows: the first section, *Introduction*, provides discussion and theoretical background. The second section, *Procedure*, describes the procedures the students must follow to complete the experiment successfully. These two sections should be carefully studied before beginning the experiment. In the third section, *Data Analysis,* the students learn how to draw conclusions from data obtained. Section four, *Error Analysis,* is unique to this volume. Students are asked to evaluate the likely magnitude of error in the data of final computed results from various sources of experimental inaccuracy. The purpose of this exercise is to impress upon students that not all parts of an experiment are equivalent and that some measurements are more critical to a successful experiment than others. At the end of each experiment is a set of *Self-study questions* to aid students in mastering the material.

All experiments, where feasible, are written with the idea that students are issued an unknown whose properties or composition are to be determined. We have found that this procedure encourages the most careful work on the part of students. Local conditions may make it appropriate to substitute knowns for unknowns in some cases.

S.E.W. and J.J.L.

CONTENTS

I

INTRODUCTORY EXPERIMENTS

EXPERIMENT

The Density of a Liquid: An Introduction to Weighing

I. INTRODUCTION

Of the physical properties that characterize pure substances, density is one that can usually be determined rapidly and accurately. The densities of a mixture of liquids may yield considerable information about the intermolecular forces existing between the molecules in the mixture when considered in conjunction with energies of vaporization. Only occasionally are the densities of pure substances additive when the substances are mixed, and this is observed principally when the effective molecular volumes of the components are very nearly equal. In some cases the volume of a solution is actually *less* than the volume of either component alone. Yet information concerning the densities of mixtures is often used as an analytical tool to establish the composition of the liquid. A familiar example is the specific-gravity measurement used to determine the concentration of sulfuric acid solution in a lead storage battery.

In this experiment you will determine the composition of a salt-water mixture by determining its density from the weight of a known volume of the sample. The volume of the sample will be measured with a pipette and the weight obtained with an analytical balance. First you will have to calibrate the volume of your pipette with distilled water.

II. PROCEDURE

A. Calibration of Pipette

If you have already calibrated a 5 ml pipette, use it to determine the density of your unknown solution. If you have not, calibrate a 5 ml pipette according to the following directions. Weigh an empty 50 ml beaker on an analytical balance. Measure 5 ml of distilled water with your pipette and place it directly into the pre-

Table 1.1
The Absolute Density of Water at
Various Temperatures

Temperature °C	Density gr/cm³	Temperature °C	Density gr/cm³
17.0	0.998774	22.0	0.997770
17.2	0.998739	22.2	0.997724
17.4	0.998704	22.4	0.997678
17.6	0.998668	22.6	0.997632
17.8	0.998632	22.8	0.997585
18.0	0.998595	23.0	0.997538
18.2	0.998558	23.2	0.997490
18.4	0.998520	23.4	0.997442
18.6	0.998482	23.6	0.997394
18.8	0.998444	23.8	0.997345
19.0	0.998405	24.0	0.997296
19.2	0.998365	24.2	0.997246
19.4	0.998325	24.4	0.997196
19.6	0.998285	24.6	0.997146
19.8	0.998244	24.8	0.997095
20.0	0.998203	25.0	0.997044
20.2	0.998162		
20.4	0.998120		
20.6	0.998078		
20.8	0.998035		
21.0	0.997992		
21.2	0.997948		
21.4	0.997904		
21.6	0.997860		
21.8	0.997815		

weighed flask.* Weigh the flask plus water using the analytical balance. The difference in the weights represents the weight of water delivered by the pipette. Density is the weight of a sample divided by its volume (Eq. 1).

$$D = \frac{W}{V} \tag{1}$$

Therefore, if we know the density, we can determine the volume of water delivered from the weight of the sample by using Equation 2.

$$V = \frac{W}{D} \tag{2}$$

Although the density of water varies with temperature, accurate values of density at various temperatures are available (see Table 1.1). Measure the temperature of the distilled water you used in this part of the experiment and use the correspond-

---

*Your laboratory instructor will demonstrate the proper use of a pipette, which is also described in Appendix 1.2.

ing value of the density of water from Table 1.1 to calculate the volume of water delivered by your pipette.

You should do at least 3 independent experiments to determine the volume of water delivered by your pipette. Take the average value as the "best value." Arrange the data in your notebook using a table containing the following information. This calibrated pipette should be used for all subsequent experiments.

B. The Density of a Salt-Water Mixture

Using the same general procedure as described in section A, determine the density of an unknown mixture of salt and water. For the volume of the solution delivered from your 5 ml pipette use the volume determined from your calibration experiments.

Be sure your pipette is clean and that it is rinsed with a small portion of your unknown solution. The latter is important to ensure that residual water clinging to the walls does not dilute the unknown solution.

Three determinations should be made and the data entered in your notebook in a tabular form similar to Table 1.3, p. 8. Report the "best value" of the density as the average value of at least three determinations.

III. DATA ANALYSIS

The composition of salt-water mixtures is directly related to the densities of these solutions (see Table 1.2). You should be able to estimate the composition of your solution with the use of Table 1.2, by plotting the composition against the density and then reading off from this graph the composition which corresponds to your experimental density.

Table 1.2
Density of Aqueous Sodium Chloride Solutions

Percent by Weight in Soln	0°C	10°C	20°C	25°C	30°C	40°C	50°C	60°C	80°C	100°C
1	1.00747	1.00707	1.00534	1.00409	1.00261	0.99908	0.99482	0.9900	0.9785	0.9651
2	1.01509	1.01442	1.01246	1.01112	1.00957	1.00593	1.00161	0.9967	0.9852	0.9719
4	1.03038	1.02920	1.02680	1.02530	1.02361	1.01977	1.01531	1.0103	0.9988	0.9855
6	1.04575	1.04408	1.04127	1.03963	1.03781	1.03378	1.02919	1.0241	1.0125	0.9994
8	1.06121	1.05907	1.05589	1.05412	1.05219	1.04798	1.04326	1.0381	1.0264	1.0134
10	1.07677	1.07419	1.07068	1.06879	1.06676	1.06238	1.05753	1.0523	1.0405	1.0276
12	1.09244	1.08946	1.08566	1.08365	1.08153	1.07699	1.07202	1.0667	1.0549	1.0420
14	1.10824	1.10491	1.10085	1.09872	1.09651	1.09182	1.08674	1.0813	1.0694	1.0565
16	1.12418	1.12056	1.11621	1.11401	1.11171	1.10688	1.10170	1.0962	1.0842	1.0713
18	1.14031	1.13643	1.13190	1.12954	1.12715	1.12218	1.11691	1.1113	1.0993	1.0864
20	1.15663	1.15254	1.14779	1.14533	1.14285	1.13774	1.13238	1.1268	1.1146	1.1017
22	1.17318	1.16891	1.16395	1.16140	1.15883	1.15358	1.14812	1.1425	1.1303	1.1172
24	1.18999	1.18557	1.18040	1.17776	1.17511	1.16971	1.16414	1.1584	1.1463	1.1331
26	1.20709	1.20254	1.19717	1.19443	1.19170	1.18614	1.18045	1.1747	1.1626	1.1492

Your report on this experiment should include the following:
1. the calibration volume of your pipette;
2. the density of your unknown salt-water solution;
3. the percent-composition of your salt-water solution;
4. the mole fraction of salt in your unknown;
5. the molarity of salt in your unknown.

For all multiple determinations, calculate the standard deviation (Appendix 1.1) for the best value.

IV. ERROR ANALYSIS

Calculate the maximum uncertainty in the volume of your pipette that arises from the following sources:

1. an error in the temperature of the water of $\pm 0.2°C$ used to calibrate the pipette;

2. use of a wet beaker to collect the calibrating water;

3. an error of ± 0.005 g in the initial weight of the beaker;

4. an error in the temperature of the water of ±0.2°C used to calibrate your 5 ml pipette;

5. an error in the temperature reading of ±0.2°C in the temperature of the salt solution. Assume the pipette was calibrated correctly.

NAME   **DATE**

SECTION

Table 1.3
Calibration Data for 5 ml Pipette

Sample Number	Weight of Flask	Weight of Flask and Water	Weight of Water	Calculated Volume of Water
1				
2				
3				
4				
			Average	

NAME   DATE

SECTION

SELF-STUDY QUESTIONS

1. In terms of the kinetic molecular theory, explain why liquids are generally less dense than solids.

2. Consider a mixture prepared by taking equal volumes of two pure substances of densities X and Y. What can be said about the density of the mixture?

3. Assume that you obtained 5.08, 5.01, 5.10 ml for the volume of your pipette after three calibration experiments. What are (a) the best value for the volume of this pipette and (b) its standard deviation?

4. Assume the pipette described in problem 3 was used to deliver a sample of a salt solution and the sample weighed 5.4971 g. What is the density of the salt solution?

5. The density of a 6 percent salt solution is 1.04127 g/ml at 20°C. What is the mole fraction of salt in this solution?

6. Calculate the molarity of the salt solution described in problem 5.

7. Using the data in Table 1.2, estimate
a. the density of a 2.25 percent salt solution at 20°C.

b. the density of a 4 percent salt solution at 45°C.

c. the density of a 5 percent salt solution at 33°C.

8. What is the weight of 24°C water delivered by the pipette described in problem 3?

APPENDIX 1.1. ANALYSIS OF THE PRECISION OF AN EXPERIMENT

When multiple results are obtained on the same experimental quantity—e.g., the concentration of copper or nickel in solution, the amount of acidic acid present in a solution—it is important to give the *best value* of the results, often taken as a simple average, together with a measure of the reproducibility of the individual determinations. The latter is often expressed as the standard deviation of the results.

The standard deviation is defined as follows:

$$\text{standard deviation} \equiv \sigma = \left[\frac{\sum_{i=1}^{N} (X_i - \bar{X})^2}{N - 1} \right]^{\frac{1}{2}}$$

where X_i = the ith data point
   N = the number of data points
   $\bar{X}$ = the average value of all the X_i's,

$$\bar{X} = \frac{\sum_{i=1}^{N} X_i}{N}$$

Note that σ is a measure of the *reproducibility* of individual experiments. It is often convenient to calculate the percentage standard deviation,

$$\text{percentage standard deviation} = \frac{\sigma}{\bar{X}} \times 100$$

The standard deviation of the mean is defined as:

$$\text{standard deviation of the mean} = \bar{\sigma} = \frac{\sigma}{\sqrt{N}}$$

Since the value of σ tends to be constant for a given experiment, the standard deviation of the mean tends to become smaller as N increases, which means that the precision of the final answer tends to be greater. One generally quotes the final result of an experiment as $\bar{X} \pm \bar{\sigma}$, where it is expected that the "true" answer will lie in the range $X - \sigma$ to $X + \sigma$. This expectation will be realized only if one has *random errors*. A systematic error (i.e., an error made throughout the experiment) invalidates this expectation.

Example: Suppose a beaker was weighed on 3 different occasions and the results 12.1806 g, 12.1821 g, and 12.1815 g were obtained. The average weight would be:

$$\frac{12.1806 + 12.1821 + 12.1815 \text{ g}}{3} = 12.1814 \text{ g}.$$

The standard deviation would be:

$$\sigma = \left\{ \frac{1}{3-1} \left[(12.1806 - 12.1814)^2 + (12.1821 - 12.1814)^2 \right. \right.$$
$$\left. \left. + (12.1810 - 12.1814)^2 \right] \right\}^{1/2}$$

$$= \{ {}^{1}\!/_{2} [(-.0008)^2 + (.0007)^2 + (.0001)^2] \}^{1/2}$$

$$= \left\{ \frac{1.14 \times 10^{-6}}{2} \right\}^{1/2} = .0007^5 \sim .0008 \text{ g.}$$

The percentation deviation would be $(.0008/12.1814) \times 100 = .0066\%$.
  The standard deviation of the mean would be:

$$\bar{\sigma} = \frac{.0008}{\sqrt{3}} = .0004^6 \sim .0005,$$

such that one would quote the final result for the weight of the beaker as

$$12.1814 \pm .0005 \text{ g.}$$

If, however, the beaker were not completely clean, this weight would still not represent the correct intrinsic weight of the beaker. This would then be an example of a systematic error.

EXPERIMENT

2

The Density of a Metal: An Introduction to Least-Squares Analysis

I. INTRODUCTION

Density is a physical property of considerable importance in characterizing a solid substance. It is often possible to identify a pure metal on the basis of its density (see Table 2.1).

II. PROCEDURE

The density of a granular metal sample will be determined from Archimedes' principle, *viz.*, from the volume of water a given weight of metal displaces. The ap-

Table 2.1
Densities of Some Common Metals

Metal	Density g/cm³
Sodium	0.97
Magnesium	1.74
Aluminum	2.70
Barium	3.78
Antimony	6.62
Chromium	6.92
Zinc	7.04
Tin	7.30
Iron	7.85
Cadmium	8.54
Nickel	8.60
Copper	8.93
Bismuth	9.67
Lead	11.34
Gold	19.30
Platinum	21.37

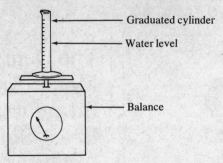

Fig. 2.1 *A diagrammatic representation of the apparatus required to determine the density of a solid using Archimedes' principle.*

paratus consisting of a 10 ml graduated cylinder is placed on a balance (Fig. 2.1). A given volume of water is placed in the cylinder, the level of water noted, and increments of the unknown metal added. After each increment the weight of the system (cylinder-water-metal) and the level of the water in the cylinder are recorded.

III. DATA ANALYSIS

From the known relationship of density (D) to volume (V) and weight (W) (Eq. 1)

$$D = \frac{W}{V}, \tag{1}$$

the data collected can be treated to give the density of the metal. The "best value" for density can be obtained from the data collected for each increment, and all the values averaged; or the data can be treated analytically by recognizing that Equation 1 can be rearranged to give Equation 2.

$$W = DV \tag{2}$$

If the weight of each increment is plotted against its volume, a straight line with slope D should result. The best straight line through the data can be obtained from a least-squares fit.

Least-Squares Analysis

There are many experimental situations in physics, chemistry, and biology where a set of experimental data, according to theory, should obey the general equation of a straight line as shown in Equation 3.

$$Y = mX + b \tag{3}$$

where m is the slope of the line and b is the Y intercept. Indeed, you will have several occasions to use the least-squares techniques in your chemistry courses, so you should learn about the idiosyncrasies of the method as early in your studies as possible. The least-squares method of fitting experimental points to a theoretical equation is one of the most often used techniques in all of the experimental sciences. In this discussion we shall limit ourselves to the straight-line type theoretical equation (Eq. 3), the simplest case of a large set of equation types.

The least-squares technique is easiest to apply when one of the experimental variables is known with a high degree of accuracy and the significant error is restricted to the other variable. See Appendix 2 for a derivation of the method. Let us consider the relationship between weight and volume given by the expression of density (Eq. 2). Figure 2.2 shows the theoretical straight line one would expect for the relationship between the weight and volume of a substance which has a density of 2 g/ml. The general equation for the straight line $Y = mX + b$ becomes $W = DV$ for this specific case where the Y-variable becomes W, the slope $m = D = 2$, and the X-variable is V and $b = 0$.

Suppose we go into the laboratory to test this relationship. Because of experimental errors (reading the balance, error in the balance, etc.) the data points will scatter about the theoretical straight line as shown in Figure 2.3.

Let us now consider a similar system except this time we don't know the density and we are not sure that the balance is properly calibrated (i.e., that a zero reading is really zero). This time we cannot draw a straight line using the theoretical equation since we don't know D. We can, however, collect experimental points and attempt to draw a straight line through them. Our problem is to draw the best straight line through the points by using the least-squares method, which defines the best line as one in which the sum of the deviations squared at each point from the line is minimized; symbolically the sum of the deviations squared is given by

$$\Sigma \, d_i^2. \tag{4}$$

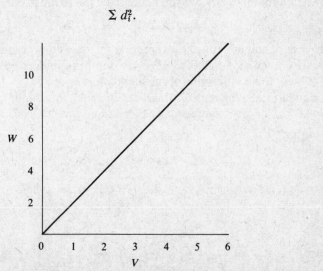

Fig. 2.2 *The equation (Eq. 2) for the relationship between weight and volume yields a straight line which goes through the origin. That is, the intercept b is zero and W and V both go to zero in a proportional manner.*

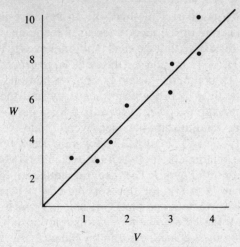

Fig. 2.3 *The experimental points often do not lie on the theoretical curve. In this figure the solid line represents the expected theoretical behavior and the points represent typical experimental data.*

For the sake of developing a generalized mathematical concept let us return to the general equation of a straight line (Eq. 3), remembering that Y corresponds to W, X corresponds to V, m corresponds to D, and b is the Y intercept. In our specific case we must consider that b may not be zero, for if the zero reading on the balance is not really zero, then b will not be zero. In addition b may not be zero because the weight plotted is the *total* weight of the system (which includes the graduated cylinder) and not only the weight of the sample.

The least-squares principle requires that the sum of the squares of the deviations be a minimum; that is, we must minimize the expression:

$$\Sigma \, d_i^2 = \Sigma (Y_i - mX_i - b)^2. \tag{5}$$

The process of minimizing the expression in Equation 3 requires the use of calculus and is given in Appendix 2. We are primarily interested in using the results of this process to find the value of the parameters m and b.

The equations for m and b which result from the minimization are given as Equations 6 and 7. The least-squares technique has its basis in statistical theory

$$m = \frac{n \sum_{i=1}^{n} X_i Y_i - \sum_{i=1}^{n} X_i \sum_{i=1}^{n} Y_i}{n \sum_{i=1}^{n} X_i^2 - \left(\sum_{i=1}^{n} X_i\right)^2} \tag{6}$$

$$b = \frac{\sum_{i=1}^{n} X_i^2 \sum_{i=1}^{n} Y_i - \sum_{i=1}^{n} X_i \sum_{i=1}^{n} X_i Y_i}{n \sum_{i=1}^{n} X_i^2 - \left(\sum_{i=1}^{n} X_i\right)^2} \tag{7}$$

which says that the least-squares method will give the most probable values of m and b if the deviations (errors) are normally distributed. The validity of the least-squares method is also restricted by the number of data points used in the analysis—that is, the more data points, the more valid the estimates of m and b.

Plot the weight of the system (cylinder and water plus increment of metal added) against the volume reading of the graduated cylinder. Calculate the slope and intercept of the best straight line using Equations 6 and 7 and draw the best line through the points using these values.

The report should contain the table of experimental data, the plot of weight vs. volume with the least-squares straight line drawn in, and the calculations leading to the slope and intercept using the least-squares analysis. Report also the density of the metal and its identity.

IV. ERROR ANALYSIS

Estimate the uncertainty in D that arises from the following sources.

1. The balance zero is in error by $+0.01$ g; i.e., the balance reads $+0.01$ when it should read zero.

2. The initial volume of the water in the graduated cylinder is misread by -0.2 ml.

3. The quantity ΣX_i^2 in Equations 6 and 7 is miscalculated as $(\Sigma X_i)^2$.

4. The sample of metal is wet with water.

NAME   **DATE**

SECTION

Table 2.2
Least-Squares Analysis

Increment i	Weight w_i	Volume v_i	$(w_i)^2$	$(v_i)^2$	$w_i \times v_i$
Initial values $i = 1$					
	$\Sigma(w_i) =$	$\Sigma(v_i) =$	$\Sigma(w_i)^2 =$	$\Sigma(v_i)^2 =$	$\Sigma w_i v_i =$

[From Eq. (6)] $m =$
[From Eq. (7)] $b =$

NAME   DATE

SECTION

SELF-STUDY QUESTIONS

1. Why is density a quantity that is characteristic of a metal?

2. What does the symbol Σ mean in Eqs. 6 and 7?

3. Assume that the following data were collected in an Archimedes-type experiment using a sample consisting of metal shot

incremental weight, g	incremental volume, ml
0.24	0.06
1.01	0.30
1.23	0.38
1.30	0.40
1.59	0.51

Give the values of the following expressions.
  a. n e. ΣX_i^2
  b. $\Sigma X_i Y_i$ f. $(\Sigma X_i)^2$
  c. ΣX_i
  d. ΣY_i

4. From the data in question 3, determine the "best value" for the density of the metal.

5. What is the value of the intercept for the least-squares analyses for the data in question 3?

6. Assume that an aqueous solution of HC1 was used accidentally instead of water as the displacement liquid in an Archimedes-type experiment. Discuss the error expected in the density of a metal assuming that this reagent (a) does not react with the metal, (b) reacts slowly to form a soluble product, (c) reacts slowly to form an insoluble product that eventually coats the metal.

APPENDIX 2. DERIVATION OF LEAST-SQUARES SOLUTION FOR SLOPE AND INTERCEPT

Given that there is a set of X_i, Y_i data which should obey an equation of the form

$$Y = mX + b \tag{1}$$

which defines a straight line of slope m and Y-intercept of b, the determination of the slope and intercept of a straight line which best fits the $X_i Y_i$ data may be performed with the aid of the least-squares principle.

The least-squares procedure is most straightforward and easy to apply when one of the experimental variables (X in this case) is known with a high degree of accuracy and the significant error is restricted to the other variable (Y in this case).

Consider Figure 2.4. In this figure, the vertical distance (d_i) between a particular point and the least-squares best-fitting line is a measure of the error associated with the point.

At the X_i coordinate Y_i is given by $mX_i + b$ according to Equation 1; hence, d_i is given by the equation:

$$d_i = Y_i - (mX_i + b) \tag{2}$$

The least-squares principle requires that the sum of the squares of the deviations be a minimum. That is, expression 3 must be a minimum.

$$\sum_{i=1}^{n} d_i^2 \tag{3}$$

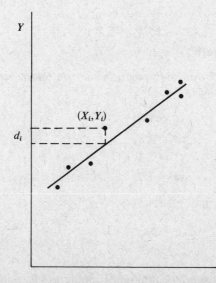

Fig. 2.4 *The relationship between the least-square line and experimental points.*

Substituting Equation 2 into expression 3 gives

$$\sum_{i=1}^{n} d_i^2 = \sum_{i=1}^{n} (Y_i - mX_i - b)^2. \tag{4}$$

Since the data points (X_i, Y_i) are all numbers, only two parameters (M and b) need to be estimated by minimizing the sums.

According to the procedures of calculus, we must differentiate Equation 4 with respect to both m and b and set the results equal to zero. First, with respect to m, we obtain:

$$\frac{\partial}{\partial m}\left(\sum_{i=1}^{n}(Y_i - mX_i - b)^2\right) = 2\sum_{i=1}^{n}(Y_i - mX_i - b)(-X_i) \tag{5}$$

$$= -2\sum_{i=1}^{n}(X_iY_i - 2m\Sigma X_i^2 - 2b\Sigma X_i = 0. \tag{6}$$

Then, differentiating Equation 4 with respect to b, we obtain:

$$\frac{\partial}{\partial b}\left(\sum_{i=1}^{n}(Y_i - mX_i - b)^2\right) = 2\sum_{i=1}^{n}(Y_i - mX_i - b)(-1) \tag{7}$$

$$= -2\sum_{i=1}^{n}Y_i + 2m\sum_{i=1}^{n}X_i + 2nb = 0. \tag{8}$$

Factoring out the common factor of 2 in Equations 6 and 8 rearranging both equations, we obtain:

$$m(\Sigma X_i^2) + b\Sigma X_i = \Sigma X_iY_i \tag{9}$$

$$m(\Sigma X_i) + b(n) = \Sigma Y_i \tag{10}$$

For the sake of ease of algebraic manipulation and notation, let us make the following definitions:

$$s1 = \Sigma X_i^2 \tag{11}$$

$$s2 = \Sigma X_i \tag{12}$$

$$s3 = \Sigma X_iY_i \tag{13}$$

$$s4 = \Sigma Y_i \tag{14}$$

and rewrite Equation 9 and Equation 10 as:

$$ms1 + bs2 = s3 \tag{15}$$

$$ms2 + b(n) = s4. \tag{16}$$

To solve Equations 15 and 16 for m, we multiply Equation 15 by n and Equation 16 by $s2$, and then subtract. The resulting equations are

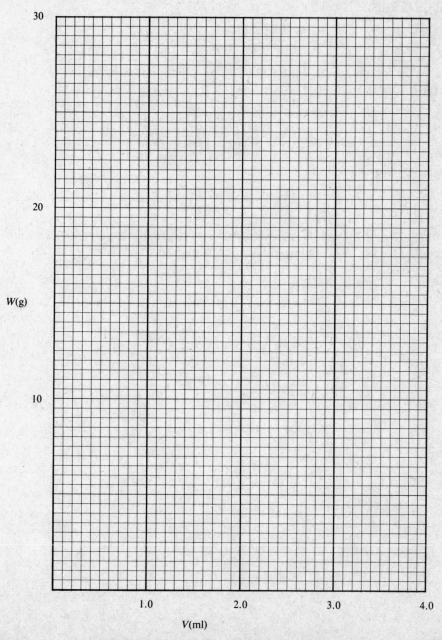

Fig. 2.5 *Graph for data, least-square line, for Experiment 2.*

$$nms1 + bns2 = ns3 \tag{17}$$
$$-(m(s2)^2 + bns2 = s2s4) \tag{18}$$
$$\overline{nms1 - m(s2)^2 = ns3 - s2s4.} \tag{19}$$

Equation 19 can be solved for m to give

$$m = \frac{ns3 - s2s4}{ns1 - (s2)^2} = \frac{n\Sigma X_i Y_i - \Sigma X_i \Sigma Y_i}{n\Sigma X_i^2 - (\Sigma X_i)^2}. \tag{20}$$

To solve for b, multiply Equation 15 by $s2$ and Equation 16 by $s1$, and then subtract the resulting equations.

$$ms1s2 + b(n)s1 = s1s4 \tag{21}$$
$$-ms1s2 + b(s2)^2 = s2s3 \tag{22}$$
$$\overline{b(n)s1 - b(s2)^2 = s1s4 - s2s3} \tag{23}$$

Equation 24 can be solved for b to give

$$b = \frac{s1s4 - s2s3}{ns1 - (s2)^2} = \frac{\Sigma X_i^2 \Sigma Y_i - \Sigma X_i \Sigma X_i Y_i}{n\Sigma X_i^2 - (\Sigma X_i)^2}. \tag{24}$$

Equations 23 and 24 give the mathematical procedure of obtaining the slope and intercept of the best straight line obtained from experimental data.

EXPERIMENT

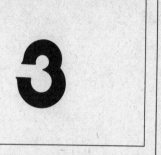

Preparation of PbCl₂(s) from Pb(C₂H₃O₂)₂: An Introduction to Inorganic Synthesis

I. INTRODUCTION

It often occurs that different salts of the same cation will have vastly different solubilities and this fact is used to advantage in qualitative analysis or purification procedures. In this experiment, $PbCl_2$(s) will be precipitated from a solution of $Pb(C_2H_3O_2)_2$. The resulting $PbCl_2$(s) will contain a significant amount of absorbed impurity ions, and recrystallization will be necessary to purify the product $PbCl_2$. This experiment represents a particularly straightforward example of inorganic analysis.

II. PROCEDURE

The reaction involved is given by Equation 1.

$$Pb^{2+} + 2Cl^- \longrightarrow PbCl_2(s) \quad \text{(white)} \tag{1}$$

The Pb^{2+} ion comes from $Pb(C_2H_3O_2)_2$, which is soluble in water. Either HCl or NaCl can provide the Cl^- ions. Lead chloride is insoluble under the conditions of the experiment. Acetate ions, $C_2H_3O_2^-$, and the counter ion accompanying the Cl^- ion, either H^+ or Na^+, remains in solution. The solid $PbCl_2$ is washed with cold water, which removes most of the soluble impurities absorbed on the surface of the product. The solid can be further purified by recrystallization; in this process the product will be partially dissolved in boiling water, quickly filtered, and allowed to cool slowly. The resulting crystals will be considerably purer than the original precipitate because of the property crystals have of excluding from the crystal lattice any ions with significantly different sizes and charges from those that make up the bulk of the crystal.

1. Weigh out 4.5–5.0 g of $Pb(C_2H_3O_2)_2$ and record the weight to the nearest mg.

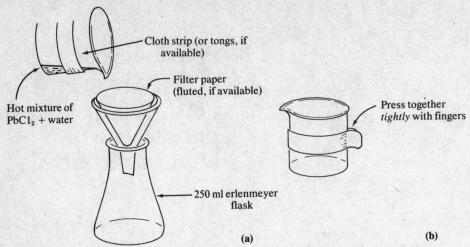

Cloth strip (or tongs, if available)

Filter paper (fluted, if available)

Hot mixture of PbCl$_2$ + water

Press together *tightly* with fingers

250 ml erlenmeyer flask

(a)   **(b)**

Fig. 3.1 **(a)** *Typical filtration arrangement.* **(b)** *Method for handling a hot beaker.*

Dissolve in approximately 100 ml of water and filter, catching the filtrate in a 250 ml beaker.

2. Calculate the amount of 3N HCl that should be added to react with the amount of Pb(C$_2$H$_3$O$_2$)$_2$ in solution. Add that amount of 3N HCl, *slowly and with stirring*. Then add 1 ml of 3N HCl in excess. After the white precipitate has settled, decant the supernatent solution and discard. Add 100 ml of distilled water for the next step, recrystallization.

3. Heat the PbCl$_2$ and water mixture to boiling and boil gently until most of the solid has dissolved. Turn off the flame, allow any remaining solid to settle (wait only 15–30 sec), then decant the hot liquid through filter paper, using an arrangement like Figure 3.1(a).

> ☞ **CAUTION:** Handle the hot beaker with a strip of cloth (Fig. 3.1(b)) or as instructed by the laboratory supervisor.

Solid material will probably appear immediately in the hot filtrate, but the filtrate may be reheated gently until the solid PbCl$_2$ redissolves. Allow the hot filtrate to cool slowly without stirring. If a significant amount of solid PbCl$_2$ remains in the 250 ml beaker, more water may be added and heated until all the remaining PbCl$_2$ dissolves.

NOTE: Add 20 ml of water initially, then in 5 ml amounts, to avoid adding a large excess of water.

This hot solution may be filtered like the first batch, with the second filtrate being combined with the first filtrate. Reheat the filtrate until essentially all solid redissolves, then allow to cool slowly, without stirring. Store the filtrate and crystals until the next laboratory period.

4. The crystals of the PbCl$_2$ will be small, white needles. Decant most of the

liquid, filter these crystals from the remaining solution (same set-up as Fig. 3.1(a)) and wash with a small amount of 95 percent ethanol. Spread out the filter paper and allow the crystals to air-dry. Alternatively the filter paper and funnel can be used with an aspirator and air drawn through the solid. The air-dried crystals may be scraped into a dry, clean, 100 or 50 ml beaker that has been previously weighed and briefly dried in an oven. The weight of the PbCl$_2$(s) can then be determined, and the yield calculated according to Equation 2:

$$\text{Yield} = \frac{\text{number of grams PbCl}_2 \text{ recovered}}{\text{number of grams PbCl}_2 \text{ theoretically possible}} \times 100. \tag{2}$$

NOTE: Retain the purified PbCl$_2$(s), which may be used for another experiment.

III. ERROR ANALYSIS

Determine the effect on the yield of the synthesis of PbCl$_2$ that arises from the following sources:

1. The reprecipitated PbCl$_2$ is filtered while the solution is 40°C rather than room temperature.

2. The recrystallized PbCl$_2$ is washed with hot water rather than 95 percent ethanol.

3. The reprecipitated PbCl$_2$ is not washed with 95 percent ethanol and is weighed immediately after filtration.

4. The molecular weight of Pb(C$_2$H$_3$O$_2$)$_2$ is taken as 266.19.

5. A 1.5 M solution of HCl is used by mistake instead of the 3 M solution called for.

1. Data

   grams of $Pb(C_2H_3O_2)_2$ _____

   grams of $PbCl_2$ recovered _____

   theoretical yield of $PbCl_2$ _____

   % yield $= \dfrac{\text{gram of } PbCl_2 \text{ recovered}}{\text{theoretical yield of } PbCl_2} \times 100 =$ _____

2. Calculation of theoretical yield

   theoretical yield of $PbCl_2$
   $$= \text{gram of } Pb(C_2H_3O_2)_2 \times \text{ratio of gram molecular weights}$$

3. Results:

NAME   **DATE**

SECTION

SELF-STUDY QUESTIONS

1. Write a balanced equation for the preparation of lead chloride from lead acetate and sodium chloride in aqueous solution.

2. What would be the product present if, after all the $PbCl_2$ were removed by filtration, the filtrate were evaporated to remove all the water? Write the chemical formula for this product.

3. What is the maximum weight of $PbCl_2$ that can be prepared from 5.12 g of lead acetate?

4. If the yield for the reaction given in problem 4 is 82 percent, what weight of $PbCl_2$ would you expect to be able to isolate?

5. Assume that 2.5 M HCl was used to prepare the $PbCl_2$ described in problem 3. What volume of HCl solution would be required?

6. Assume that pure NaCl was used to prepare the $PbCl_2$ described in problem 3. What weight of NaCl would be required?

7. What weight of sodium acetate would you expect to find in solution in the experiment described in problem 6?

EXPERIMENT

4

Analysis of Water in Hydrates: An Introduction to Gravimetric Analysis

I. INTRODUCTION

Gravimetric analysis is one of the most useful methods of analytical chemistry and is based upon the law of constant proportions. Generally, gravimetric analysis involves (a) weighing a sample, (b) treating the sample to obtain the component of interest in a form that can be weighed, and (c) weighing that substance.

In this experiment we shall analyze an inorganic hydrate for its water content. When hydrates are heated, water is driven off (Eq. 1), leaving

$$\text{hydrate} \xrightarrow{\text{heat}} \text{anhydrous compound} + H_2O \tag{1}$$

the corresponding anhydrous substance. If the hydrate and the anhydrous compound are non-volatile, the difference between the weight of the hydrate and the anhydrous compound is the weight of water present in the hydrate.

II. PROCEDURE

Wash and dry a crucible. Heat it with a bunsen burner for about 30 seconds (Fig. 4.1).

Allow the crucible to cool in place. Using tongs, place the cool crucible on a clean, dry evaporating dish, carry it to the analytical balance, weigh and record the weight of the crucible. Place about 4 g of your unknown hydrate in the crucible, weigh and record the weight.

Replace the crucible and its contents on the clay triangle and heat the bottom of the crucible over a small flame; pass the flame around the bottom until the crucible and its contents are evenly heated. Then heat the crucible more strongly, but do not leave the burner with a high flame unattended under the crucible. Be careful not to heat the contents so strongly that the contents pop and scatter from the crucible.

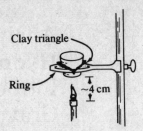

Fig. 4.1 *Method for heating a crucible with a bunsen burner.*

Let the crucible cool on the triangle to room temperature, handle the crucible with tongs, place it in the evaporating dish, carry it to the balance room and weigh it. Record the weight.

To ensure that all the water has been removed from the hydrate, reheat the crucible containing the residue for 10 more minutes, cool, and weigh. If all the water was removed in the original heatings the weights should be the same within experimental error. If the second weight is not the same as the first weight, reheat the crucible containing the residue for 10 minutes more, cool, and weigh. This procedure should be continued until the residue comes to constant weight. You will usually require only one or two additional periods of heating to bring the crucible to constant weight.

Repeat the experiment two more times.

III. DATA ANALYSIS

The difference in weight of the sample before and after heating corresponds to the weight of water driven off. Thus the percentage of water present in the hydrate can be calculated in the usual way (Eq. 2).

$$\frac{\text{weight of } H_2O}{\text{weight of hydrate}} \times 100 = \text{percent } H_2O \tag{2}$$

Report the average value of the three determinations and the standard deviation for your experimental results.

IV. ERROR ANALYSIS

Determine the effect on the analysis of a hydrate that arises from the following sources:
   1. The weight of the empty crucible is in error by -0.005 g.

2. The first final weight of the crucible after the first heating is 24.9131 g and after the second heating it is 24.8876 g.

3. A small particle of hydrate pops out of the crucible because it was heated too strongly.

4. The zero point of the balance is really +0.0100 g rather than 0.0000; this value does not change during the course of the analysis.

5. It is discovered that the residue is hygroscopic. The heated crucible is kept in a beaker in your drawer until next period before it is weighed finally.

1. Table of Data

	Trial		
	1	2	3
w_1, weight of clean, dry crucible			
w_2, weight of crucible + hydrate			
w_3, weight of cool crucible + hydrate after first heating			
w_4, weight of cool crucible + hydrate after second heating			
$(\% \ H_2O)_1 = \dfrac{w_2 - w_3}{w_2 - w_1} \times 100$			
$(\% \ H_2O)_2 = \dfrac{w_2 - w_4}{w_2 - w_1} \times 100$			
$(\% \ H_2O)_1$ average		$\sigma =$	
$(\% \ H_2O)_2$ average		$\sigma =$	

2. Calculation of standard deviation

NAME DATE

SECTION

SELF-STUDY QUESTIONS

1. What is the percent of water present in the hydrate $CuSO_4 \cdot 5H_2O$?

2. If 5.6318 g of $CoCl_2 \cdot 6H_2O$ is heated, what weight of residue would be expected?

3. Assume the following data were obtained for the hydrate $MgCl_2 \cdot XH_2O$:

weight of empty crucible = 38.6319 g;
weight of crucible plus sample = 39.6480 g;
constant weight value of crucible plus residue = 39.1080 g.

What is the percentage of water in the hydrate?

4. What is the value of X in the formula for the hydrate described in problem 3?

5. Discuss the problems you might expect if (a) the residue was volatile and (b) the hydrate was volatile, using the method described in this experiment.

6. Some carborates decompose into metal oxides and carbon dioxide when heated. Describe a procedure to analyze limestone.

7. Would you be able to determine the amount of lime (CaO) present in a sample of limestone ($CaCO_3$) which contained other non-volatile impurities (such as Fe_2O_3)? Explain your answer.

EXPERIMENT

5

An Introduction to Analysis by Gas Evolution

I. INTRODUCTION

Many chemical reactions give gases as products, i.e., $Na + H_2O \longrightarrow NaOH + \frac{1}{2}H_2(g)$. A familiar reaction that gives off a gas, one that is not well understood, is the reaction that produces oxygen in photosynthesis. Many times analyses of gaseous products can give us just as much useful information as the analysis of a solid or liquid product. The gaseous product is sometimes the only one that is easily separable from the reaction.

In this experiment CO_2 (carbon dioxide) gas will be produced by the reaction of $NaHCO_3$ (sodium bicarbonate) with sulfuric acid.

$$NaHCO_3 + H_2SO_4 \rightleftharpoons CO_2(g) + NaHSO_4 + H_2O \qquad (1)$$

Since the volume of the CO_2 evolved can be collected and measured, a quantitative measurement can be made of the amount of $NaHCO_3$ originally present, using the stoichiometry shown in Equation 1.

Baking powder is a mixture of $NaHCO_3$ and solid, weak acids like potassium hydrogen tartrate. When the baking powder is moistened the acids ionize and the hydronium ion (H_3O^+) can react with the $NaHCO_3$ to yield CO_2 (Eq. 1).

$$KHC_4H_4O_6 + NaHCO_3 \overset{H_2O}{\rightleftharpoons} CO_2(g) + H_2O + NaKC_4H_4O_6 \qquad (2)$$

A sample of baking powder will be analyzed by the collection of CO_2 gas evolved from the $NaHCO_3$ present.

II. PROCEDURE

1. Obtain a known sample of baking powder (or $NaHCO_3$) for practice.
2. Prepare 10 ml of a 3 M H_2SO_4 solution.

☛ **CAUTION:** remember to pour the acid into the water when you do this dilution.

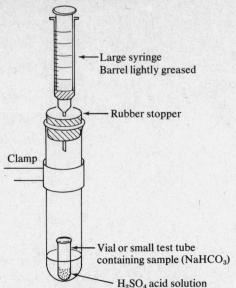

Large syringe
Barrel lightly greased

Rubber stopper

Clamp

Vial or small test tube
containing sample ($NaHCO_3$)

H_2SO_4 acid solution

Fig. 5.1 *Gasimetric analysis apparatus.*

3. Weigh by difference* an appropriate sample of baking powder into a small vial or test tube. A weight of 0.1–0.3 g of $NaHCO_3$ will produce sufficient gas to measure with a 30 ml syringe.

4. Assemble the apparatus shown in Figure 5.1. The syringe should be *lightly* greased to ensure that the barrel slides easily as the gas is evolved.

5. Tilt the apparatus until the H_2SO_4 solution runs into the vial. Repeat this procedure until no more gas is evolved. If any gas escapes from the closed system the experiment will have to be repeated.

6. Record the volume of gas obtained after the vessel cools to room temperature. Record the room temperature and the barometric pressure.

7. Obtain an unknown sample and repeat the process.

III. DATA ANALYSIS

The ideal gas law will be used to analyze the data in this experiment. It is assumed CO_2 behaves as an ideal gas. The ideal gas law states

$$PV = nRT \qquad (3)$$

where P = pressure of the gas
       V = volume of the gas
       n = number of moles of gas present
       R = universal gas constant
       T = temperature of the gas.

*See Appendix 5.

Rearrangement of Equation 3 gives:

$$\frac{PV}{T} = nR. \tag{4}$$

Therefore, if the number of moles of gas is constant during an experiment and the gas is ideal, the right side of Equation 4 under any set of experimental conditions (P, V, T) is constant. The product nR must be constant and Equation 5 must hold for the same sample of gas under two separate conditions.

$$\frac{P_1 V_1}{T_1} = \frac{P_2 V_2}{T_2} \tag{5}$$

Gases are not conveniently compared except under similar conditions of temperature and pressure. The generally accepted conditions are *standard temperature and pressure* (STP), which correspond to 273°K and 760 torr. You will recall that one mole of an ideal gas occupies a volume of 22.4 1 at STP. Corrections of gas volumes to standard conditions can be made with Equation 5.

Gases collected over aqueous solutions also contain water vapor; that is, the total pressure of this mixture (P_T) is the sum of the partial pressures (Eq. 6)

$$P_T = p_g + p_{H_2O} \tag{6}$$

where p_g and p_{H_2O} are the partial pressures of the gas and the water vapor, respectively. The rearrangement of Equation 6 to Equation 7

$$p_g = P_T - p_{H_2O} \tag{7}$$

indicates that the pressure of the dry gas can be obtained from the total pressure on the mixture (usually measured directly as the atmospheric pressure) and the vapor pressure of the water. Every liquid, whether it is pure or a solution, exerts a vapor pressure which can be measured. The vapor pressure of water and various aqueous solutions of H_2SO_4 are recorded in Tables 5.1, 5.2.

Table 5.1
Vapor Pressure of Water

Temperature, °C	Pressure, torr	Temperature, °C	Pressure, torr	Temperature, °C	Pressure, torr	Temperature, °C	Pressure, torr
0	4.580	22	19.802	32	35.629	60	149.38
5	6.536	23	21.043	33	37.695	65	187.56
10	9.197	24	22.351	34	39.863	70	233.71
15	12.771	25	23.728	35	42.139	75	289.13
16	13.617	26	25.181	36	44.527	80	355.22
17	14.511	27	26.709	40	55.288	85	433.56
18	15.457	28	28.318	45	71.840	90	525.86
19	16.456	29	30.011	50	91.492	95	634.00
20	17.512	30	31.791	55	118.03	100	760.00
21	18.626	31	33.662				

Table 5.2
Vapor Pressure[1] of H_2SO_4

%[2]	20[°3]	40°	p 60°	80°	100°
0	17.54	54.4	146.6	352	760
10	16.25	50.0	134.5	323	704
20	14.9	45.9	123.6	297	650
25	14.0	43.2	116.4	280	610
30	13.0	40.0	107.9	260	567
40	10.4	32.2	87.3	211.8	465
50	7.22	24.5	62.4	152.7	338
55	5.40	17.25	48.0	118.7	265
60	3.69	12.05	34.2	86.1	195.5
65	2.25	7.55	22.0	56.9	132.2
70	1.19	4.22	12.8	34.1	81.6
75	0.53	1.93	6.33	17.7	44.2
80	0.18	0.72	2.49	7.48	19.9
85	0.04	0.19	0.73	2.43	7.06
90	0.004	0.022	0.10	0.39	1.28
94	0.00024	0.00016	0.009	0.04	0.15
96	0.00005	0.00004	0.0024	0.012	0.05
98.3	0.00003	0.00025	0.0015	0.008	0.033
100	0.00035	0.0025	0.014	0.07	0.27

1. Pressures expressed in torr.
2. Percent by weight.
3. Temperature in °C.

As an example, assume that the following data were obtained in an experiment:

room temperature = 20°;
volume of CO_2 = 70 ml;
barometric pressure = 740 torr.

The vapor pressure of water in $3M$ H_2SO_4 solution (25%) at 20° is 14 torr (see Table 5.2), so

pressure of CO_2(dry) = 740 − 14 = 726.

Using Equation 5, correction of the volume of CO_2 to STP is made:

$$\frac{P_1V_1}{T_1} = \frac{P_2V_2}{T_2}$$

$$\frac{(726)(70 \text{ ml})}{293°} = \frac{(760)(V_2)}{273°}$$

$$V_2 = \frac{726 \times 70 \text{ ml} \times 273°}{760(293°)} = 62.3 \text{ ml.}$$

Since 1 mole of gas at STP has a volume of 22.41,

$$62.3 \text{ ml} = .0621$$

$$\frac{.062 \text{ l}}{22.4 \text{ l/mole}} = .00276 \text{ moles.}$$

An alternative method of calculating the number of moles of CO_2 formed is to use the ideal gas law (Eq. 4) solved for n.

$$n = \frac{PV}{RT} \tag{8}$$

For P expressed in torr, V in liter, and T in °K, the constant R has the value 62.363 liter torr/mole deg. Thus the number of moles for this example is given by

$$n = \frac{(722)(0.070)}{(62.363)(293)}$$

$$n = 0.00276 \text{ mole},$$

which, of course, is the same result as that obtained in the previous calculation.

And the number of grams of $NaHCO_3$ is obtained by

$$\text{number of moles} \times \text{molecular weight } \frac{g}{\text{mole}} = \text{g of } NaHCO_3$$

$$.00276 \text{ mole} \times \frac{84 \text{ g}}{\text{mole}} = .232 \text{ g of } NaHCO_3.$$

Assume that the original sample weighed 0.573 g; then the percentage of $NaHCO_3$ present is calculated by dividing the weight of $NaHCO_3$ by the weight of the sample (*both expressed in the same units*), and multiplying by 100.

$$\frac{.232g}{.573g} \times 100 = 40.5\%$$

IV. ERROR ANALYSIS

When answering the following questions, assume you have a sample of baking soda which contains 25.0 percent $NaHCO_3$. Calculate the maximum uncertainty in the percentage of $NaHCO_3$ that arises from the following sources:

1. An error of ±0.001 g is made in the weight of the sample taken for analysis.

2. The CO_2 is collected over a 10 percent solution of H_2SO_4 at 20°C but the correction for the vapor pressure of water is used in the calculations.

3. The total pressure of the gas in the syringe is in error by ±1 torr.

4. The volume of gas measured by the syringe is in error by ±1 ml.

5. The syringe barrel is not lubricated properly so that the pressure inside the syringe is higher than atmospheric pressure when the volume is determined.

6. The gas is released too rapidly, causing a momentary leak around the syringe needle, and some of the CO_2 escapes.

7. The syringe is held in the palm of your hand for 15 minutes before the volume of gas is determined.

NAME _____ - _____ DATE _____

SECTION _____

_____

Atmospheric pressure (atm or torr)_____

Room temperature (°K)_____

1. Table of data

	Trial		
	1	2	3
Weight of NaHCO$_3$ sample (w_i)			
Initial volume of syringe (v_i)			
Final volume of syringe (v_f)			
P$_{CO_2}$, corrected for vapor pressure			
Moles of CO$_2$ (eq. 8)			
Grams of NaHCO$_3$, $w_i^{NaHCO_3}$			
% NaHCO$_3$ = $\dfrac{w_i^{NaHCO_3}}{w_i} \times 100$			
Average =			
σ =			

SELF-STUDY QUESTIONS

1. Give a balanced equation for the reaction of $CaCO_3$ (limestone) with sulfuric acid.

2. Discuss the use of the reaction given in problem 1 as the basis for a method of analysis of limestone.

3. Give the characteristics of an ideal gas using the ideas of the kinetic-molecular theory.

4. Under what conditions of temperature and pressure would a gas *not* behave in an ideal manner?

5. State in mathematical form Dalton's law of partial pressure.

6. Give the equation for the ideal gas law and identify the quantities involved.

7. What volume will one mole of CO_2 occupy when collected over a 10 percent H_2SO_4 solution at 40°C?

8. Assume that a sample of baking soda weighing 0.752 g is treated with 10 percent H_2SO_4 and liberates 52 ml of CO_2, the latter being measured at 20°C at a pressure of 715 torr.

a. What volume of *dry* CO_2 (measured at STP) was liberated?

b. How many moles of $NaHCO_3$ were present in the sample of baking soda?

c. What is the percentage of $NaHCO_3$ in the baking soda?

APPENDIX 5. WEIGHING BY DIFFERENCE

The following method is the correct method for weighing by difference and should always be followed. The advantages of weighing by difference are:

1. Fewer weighings are generally required.
2. The container into which the sample is tapped does not have to be dry.
3. Since the desired weight is being obtained as the difference between two weights, there will be no inherent error if the balance was not exactly zeroed.

Instructions are given using a practical example. Assume you are instructed to weigh out a 0.1 g sample of KHP.

NOTE: It is assumed that the KHP is already dry.

PROCEDURE

1. Zero balance
2. Place total sample in small beaker and weigh sample and beaker. (Be sure you are in full weighing position before recording weight.)
3. Turn off balance.
4. Remove beaker from balance using a paper strip.

2. 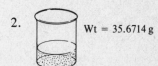 Wt = 35.6714 g

4.

5. Tap approximately 0.1 g into appropriate container.

5.

 CAUTION: Do not transfer with spatula.

6. Place beaker back on balance and reweigh.

6. Wt = 35.5231 g

EXPERIMENT

The Determination of the Equivalent Weight of an Acid: An Introduction to Titration Methods

I. INTRODUCTION

When a solution of a strong acid is mixed with a solution of a strong base, a chemical reaction occurs that can be represented by Equation 1

$$H_3O^+ + OH^- \rightleftharpoons 2H_2O \tag{1}$$

or

$$H^+ + OH^- \rightleftharpoons H_2O. \tag{2}$$

This is called a *neutralization reaction* and is used extensively to change the acidic or basic properties of solutions. The equilibrium constant for the above neutralization reaction is about 10^{14} at room temperature. Therefore the reaction can be considered to proceed to completion, using up whichever of the ions is present in the lesser amount and leaving the solution either acidic or basic, depending on whether H^+ or OH^- ion was in excess.

Since the neutralization reaction is essentially quantitative, it can be used to determine the concentrations of acidic or basic solutions. A procedure that is frequently used involves the titration of an acid with a base. In a titration a basic solution is added from a buret to a measured volume of acid solution until the number of moles of OH^- ion added is equal to the number of moles of H^+ ion present in the acid according to Equation 2; this is known as the *equivalence point of the titration*. When this occurs, the volume of basic solution that has been added should be measured. At the end point of a titration of an acid with a base,

$$\text{number of moles } H^+ \text{ originally present} = \text{number of moles } OH^- \text{ added.} \tag{3}$$

Recall the definition of the concentration unit called *molarity* (Eq. 4).

$$\text{molarity } M \text{ of species S} = \frac{\text{number of moles S in the solution}}{\text{volume of the solution in liters}} \tag{4}$$

Rearrangement of Equation 4 gives Equation 5.

number of moles S in solution = molarity of S × volume of solution in liters   (5)

The number of moles of H^+ and OH^- involved in the neutralization reaction can be calculated from Equation 5; substitution into Equation 3 gives Equation 6.

$$M_{H^+} \times V_{acid} = M_{OH^-} \times V_{base} \qquad (6)$$

Therefore, if the molarity of either the H^+ or the OH^- ion in its solution is known, the molarity of the other ion can be found from the titration.

The equivalence point or end point in the titration is determined by using a substance, called an *indicator,* that changes color at the proper pH. The indicators used in acid-base titrations are weak organic acids or bases that change color when they are neutralized. One of the most common indicators is phenolphthalein, which is colorless in acid solutions but becomes pink when the pH of the solution becomes 9 or higher.

When a solution of a strong acid is titrated with a solution of a strong base, the pH at the end point will be about 7. At the end point a drop of acid or base added to the solution will change its pH by several pH units, so that phenolphthalein can be used as an indicator in such titrations. If a weak acid is titrated with a strong base, the pH at the equivalence point is somewhat higher than 7, perhaps 8 or 9, and phenolphthalein is still a very satisfactory indicator. If, however, a solution of a weak base such as ammonia is titrated with a strong acid, the pH will be a unit or two less than 7 at the end point, and phenolphthalein will not be as good an indicator for the titration as, for example, methyl red, whose color changes from red to yellow as the pH changes from about 4 to 6. Ordinarily, indicators will be chosen so that their color change occurs at about the pH at the equivalence point of a given acid-base titration.

In this experiment you will determine the molarity of OH^- ion in an NaOH solution by titrating that solution against a standardized solution of HCl. Since in these solutions one mole of acid in solution furnishes one mole of H^+ ion and one mole of base produces one mole of OH^- ion, $M_{HCl} = M_{H^+}$ in the acid solution, and $M_{NaOH} = M_{OH^-}$ in the basic solution. Therefore the titration will allow you to find M_{NaOH} as well as M_{OH^-}. Note, however, that to be able to calculate M_{NaOH} you must know the formula of the base.

You will then use your standardized NaOH solution to titrate a sample of a pure solid organic acid. By titrating a weighed sample of unknown acid with your standardized NaOH solution, you can easily determine the number of moles of H^+ ion available in the sample. From the number of moles of H^+ and the weight of the sample you can calculate the number of grams of acid that would contain one mole of H^+ ion. This quantity is called the *gram equivalent weight* of the acid. If one mole of the acid can produce one mole of H^+, then the weight of the mole of the acid and its gram equivalent weight are equal. If, however, the acid has three moles of available H^+ ion per mole of acid, the GMW is 3 × GEW. Since you will not be given the formula of the acid, you will be able to determine only the gram equivalent weight of the acid by titration with NaOH.

II. PROCEDURE

Obtain two burets and a sample of solid unknown acid from the stockroom.

A. Preparation and Standardization of NaOH Solution

Into a small graduated cylinder, draw about 7 ml of the stock 6 M NaOH solution provided in the laboratory and dilute to about 400 ml with distilled water in a 500 ml Florence flask. Stopper the flask tightly and mix the solution thoroughly at intervals over a period of at least 15 minutes before using the solution.

Draw into a clean, dry 125 ml Erlenmeyer flask about 75 ml of standardized HCl solution (about 0.1 M) from the stock solution on the reagent shelf. This amount should provide all the standard acid you will need, so do not waste it.

Clean the two burets and rinse with distilled water. Then rinse one buret three times with a few ml of the HCl solution. Fill the buret with HCl; open the stopcock momentarily to fill the tip. Proceed to clean and fill the other buret with your NaOH solution in a similar manner. Carefully label the two burets. Check to see that your burets do not leak and that there are no air bubbles in either buret tip. Read the levels in both burets to 0.02 ml. All volumetric readings in the next paragraph should be read to 0.02 ml.

Draw about 25 ml of the HCl solution from the buret into a clean 250 ml Erlenmeyer flask; add to the flask about 25 ml of distilled H_2O and 2 or 3 drops of phenolphthalein indicator solution. Place a white sheet of paper under the flask to aid in the detection of any color change. Add the NaOH solution intermittently from its buret to the solution in the flask; note the pink phenolphthalein color that appears and disappears as the drops of base hit the acid solution and are mixed with it. Swirl the liquid in the flask gently and continuously as you add the NaOH solution. When the pink color begins to persist, slow down the rate of addition of NaOH. In the final stages of the titration add the NaOH drop by drop until the entire solution just turns a pale pink color that will persist for about 30 seconds. If you go past the end point and obtain a red solution, add a few drops of the HCl solution to remove the color, and then add NaOH a drop at a time until the pink color persists. Carefully record the final readings on the HCl and NaOH burets.

To the 250 ml Erlenmeyer flask containing the titrated solution, add about 10 ml more of the standard HCl solution. Titrate this, as before, with the NaOH to an end point and carefully record both buret readings once again. To this solution add about 10 ml more HCl and titrate a third time with NaOH.

You should have now completed three titrations, with total-HCl volumes of about 25, 35, and 45 ml. Find the ratio V_{NaOH}/V_{HCl} at the end point of each of the titrations, using total volumes of each reagent reacted up to that end point. At least two of these volume ratios should agree to within 1 percent. If they do, proceed to the next part of the experiment. If they do not, repeat these titrations until two volume ratios do agree.

B. Determination of the Gram Equivalent Weight of an Acid

Weigh the vial containing your solid acid on the analytical balance. Carefully pour out about half the sample into a clean but not necessarily dry 250 ml Erlenmeyer flask. Again weigh the vial accurately. Add about 50 ml of distilled water and 2 or 3 drops of phenolphthalein to the flask. The acid may be relatively insoluble, so don't worry if it doesn't all dissolve.

Fill your NaOH buret with your (now standardized) NaOH solution and read the level accurately.

Titrate the acid solution as before. As the acid is neutralized by the NaOH, it will tend to dissolve in the solution. If your unknown is insoluble to the extent that the phenolphthalein color appears before all the solid dissolves, add 25 ml of ethanol to the solution to increase the solubility. Record the NaOH buret reading at the end point.

Pour the rest of your acid sample into another 250 ml Erlenmeyer flask and weigh the vial accurately. Titrate the acid as before with NaOH solution. If you go past the end point in these titrations, it is possible, though more complicated in calculations, to back-titrate with a little of the standard HCl solution. Measure the volume of additional HCl used and subtract the number of moles HCl in that volume from the number of moles NaOH used in the titration. The difference will equal the number of moles NaOH used to neutralize the acid sample.

III. DATA ANALYSIS

Using Equation 6, calculate the molarity of the NaOH solution for each of your trials. Obtain the best value and estimate its standard deviation.

Now that you have standardized your NaOH solution you can use this value to estimate the equivalent weight of your unknown acid. The number of moles of base needed to neutralize the acid is calculated from Equation 5; since each mole of acid present requires one mole of base (see Eq. 2), this quantity is also the number of moles of acid present in the sample of unknown you weighed out. Thus the equivalent weight of your unknown acid is given by Equation 7.

$$\text{Eq. weight} = \frac{\text{weight of acid}}{\text{moles of H}} \tag{7}$$

IV. ERROR ANALYSIS

Calculate the maximum uncertainty in the standardization of the NaOH solution that arises from the following sources:

1. The error in a buret reading is ±0.01 ml.

2. The end point in the titration is exceeded by 0.05 ml.

Calculate the maximum uncertainty in the value of the equivalent weight of the unknown acid that arises from the following sources:

3. The error in weighing the sample titrated is ±1 mg.

4. The error in a buret reading is ±0.01 ml.

5. The error in the concentration of NaOH solution is ±0.005 moles.

NAME   **DATE**

SECTION

A. Standardization of NaOH Solution

	Trial		
	1	2	3
Final HCl buret reading			
Initial HCl buret reading			
Volume of HCl used			
Final NaOH buret reading			
Initial NaOH buret reading			
Volume of NaOH used			
Molarity NaOH solution			

Molarity HCl stock solution _____

Average molarity NaOH solution _____

Standard deviation _____

NAME   **DATE**

SECTION

B.   Equivalent Weight of Unknown Acid

	Trial	
	1	2
Weight of sample		
Volume of NaOH used		
Moles NaOH used		
Moles H^+ in sample		
Equivalent weight		

NAME   DATE

SECTION

SELF-STUDY QUESTIONS

1. Calculate the molarity of a solution prepared by diluting 7 ml of 6 *M* NaOH to a volume of 400 ml.

2. How many moles of solute are present in 75 ml of 0.0956 *M* HCl solution?

3. What is the molarity of NaOH solution if 26.95 ml are required to neutralize 22.87 ml of a 0.1034 *M* HCl solution?

4. What is the equivalent weight of each of the following acids: (a) HNO_3, (b) H_2SO_4, (c) H_3PO_4, (d) $H\,C_2H_3O_2$?

5. How many equivalent weights of acid are present in (a) 0.2 moles H_2SO_4, (b) 0.15 moles HCl, (c) 1.43 moles H_3PO_4?

6. A sample of acid (0.2531 g) required for neutralization 28.75 ml of the NaOH solution described in problem 3. What is the equivalent weight of the acid?

7. If the acid described in problem 6 is known to form a salt with the formula Na_2X, what is the molecular weight of the acid?

EXPERIMENT

7

Colorimetric Determination of Cu

I. INTRODUCTION

In the present procedure the concentration of Cu^{2+} ion in an unknown solution will be determined by comparing the color intensity of a standard Cu^{2+} solution, prepared by you, and the solution containing an unknown amount of Cu^{2+}. The formula for the species to be studied is the octahedral complex $Cu(NH_3)_4(H_2O)_2^{2+}$ and its structure is given as:

$$\left[\begin{array}{c} H\diagdown O \diagup H \\ H_3N \quad \downarrow \quad NH_3 \\ Cu \\ H_3N \quad \uparrow \quad NH_3 \\ H \diagup O \diagdown H \end{array} \right]^{2+}$$ (deep blue)

(I)

The ammonia molecules are equidistant from the Cu atom and all lie in the same plane. They are also more tightly bound to the Cu than the two water molecules. This complex ion is produced by adding an excess of NH_4OH (which contains NH_3 from the equilibrium $NH_4OH \rightleftharpoons NH_3 + H_2O$) to an aqueous solution of Cu^{2+}, which actually exists as $Cu(H_2O)_6^{2+}$ in neutral solution. The structure of $Cu(H_2O)_6^{2+}$ is similar to that shown in I except that water molecules replace the ammonia molecules in this structure; $Cu(H_2O)_6^{2+}$ is a light blue color in aqueous solution. In the course of the reaction you will notice that the blue color of the $Cu(H_2O)_6^{2+}$ ion is replaced by a much deeper blue when excess NH_4OH is added to solution.

The tetra-ammonia complex is studied rather than the $Cu(H_2O)_6^{2+}$ species in order that the colored species in solution is well defined. In the detailed procedure given in this experiment, NH_4OH is added to neutralize the acid solution formed during the preparation of the standard solution. Unless the NH_4OH is carefully removed, we will find ourselves studying a range of complexes (with slightly different colors and different intensities) in which an indeterminate number of water and ammonia molecules are attached to the Cu^{2+} ion, i.e., $Cu(H_2O)_6^{2+}$, $Cu(NH_3)$ $(H_2O)_5^{2+}$, $Cu(NH_3)_2(H_2O)_4^{2+}$, etc. Thus, we force the reaction of $Cu(H_2O)_6^{2+}$ with NH_3 to completion by adding an excess of NH_3.

The primary standard will be prepared by dissolving a known weight of copper wire in concentrated HNO_3. Nitric acid is used because Cu is insufficiently active to reduce hydrogen in an acid ($Cu^{++} + 2e = Cu$, $\Delta E_o = 0.34$). The reaction that occurs when HNO_3 is added to Cu is given by Equation 1.

$$3Cu + 2NO_3^- + 8H^+ = 3Cu^{2+} + 2NO \text{ (colorless gas)} + 4H_2O \qquad (1)$$

The colored gas that is observed arises from the further reaction of a product from Equation 1 with oxygen.

$$2NO + O_2 = 2NO_2 \text{ (red-brown gas)} \qquad (2)$$

II. PROCEDURE

Steps 1–4 deal with the preparation of the Cu^{2+} standard solution.

1. Obtain a piece of copper wire and weigh it accurately to $+0.0001$ g (use the analytical balance). If you plan to use a 250 ml volumetric flask use the small copper wire. If you plan to use a 500 ml volumetric flask use the large copper wire.

2. Put the wire in an evaporating dish or a small (50 ml) beaker and add approximately 3–5 ml of concentrated HNO_3 (under the hood). The mixture may be heated gently if the reaction proceeds slowly. Add a few more ml of HNO_3 if metallic copper is left after the gas evolution ceases. The reaction mixture may appear green at this point due to the presence of dissolved NO_2 gas. Much of this gas can be removed by stirring the solution. The solution should be blue after step 3 is completed.

3. Slowly add 10 ml of distilled water with a dropper to the evaporating dish. Add 6 M NH_4OH *dropwise* until a faint odor of ammonia is detected or until the solution is slightly basic (test with pH paper).

4. Quantitatively transfer the solution carefully to the 250 ml or the 500 ml volumetric flask. Rinse the evaporating dish or beaker with distilled water and transfer the wash water into the volumetric flask until no blue color is left behind. Add water to about half volume. If the solution looks cloudy it is due to the presence of $Cu(OH)_2$. Add dilute HNO_3 5 drops at a time to the solution and mix. When the solution is clear, add distilled water until the bottom of the meniscus rests on the etched ring on the neck of the flask. Calculate the concentration of Cu^{2+} in mg/ml for this standard solution.

The unknown solution is then compared with the known solution using the appropriate procedure described in Appendix 7.

Steps 5–8 describe the colorimetric comparison of the standard solution, prepared above, and the unknown Cu^{2+} solution.

Step 8 will be different than below if an instrumental method is used; see Appendix 7 for a description of some instrumental methods.

5. Take two test tubes (graduated test tubes will do) and label one "known" and the other "unknown." Step 6 is carried out identically for both the known and the unknown solutions.

6. Pipette 5 ml of the unknown Cu^{2+} solution into a graduated 10 ml test tube. Then add concentrated NH_4OH until the total volume is 10 ml.

7. Pipette 5 ml of the unknown Cu^{2+} solution into another graduated 10 ml test tube. Then add concentrated NH_4OH until the total volume is 10 ml.

8. Compare the color intensity of the two solutions, viewing the two test tubes lengthwise as described in Appendix 7, part E. Withdraw liquid from the more intensely colored solution (it does not matter if the more intensely colored solution is the known or unknown) until the color intensities match. Do not discard this liquid but keep in an appropriately labeled, clean, dry test tube in case you need to put it back in the viewing tube. When the color intensities match, record the solution depth of both solutions to the nearest 0.5 mm. These depths will be referred to as l_k (known) and l_u (unknown). To measure the depths, use a ruler, not the graduation marks on the test tube. Measure each several times. It is useful to withdraw some of the less intensely colored solution for subsequent determinations, so that a range of color intensities is used.

It is recommended that the comparison be made 2 or 3 times per solution pair and that the solutions be made up fresh two times (i.e., steps 6 and 7). This will lead to 5–6 independent determinations of the ratio of the solution depths at equal color intensities.

Preparing the standard Cu^{2+} solution should require no more than 45–60 minutes. This can be done well ahead of the colorimetric determination. The colorimetric determination should require no more than 60–90 minutes (the $Cu(NH_3)_4^{2+}$ solution should be discarded after use because it will lose NH_3 upon standing).

III. DATA ANALYSIS

At equal color intensity the relationship given by Equation 8 obtains. Using the concentration of the known solution, and the average value determined for the ratio l_k/l_u, the concentration of the unknown solution in mg/ml can be determined.

Suppose the sample of Cu wire weighed 0.3219 g (= 321.9 mg). When dissolved in HNO_3 and diluted to 500 ml the known solution has a concentration of $c_k = (321.9 \text{ mg}/500.0 \text{ ml}) = 0.6438$ mg/ml (note that 4 significant figures are assumed in the weighing and the volume). Suppose for a particular intensity match trial the depth of the known solution was 5.60 mm (= l_k) and the depth of the unknown solution was 6.35 mm (= l_u). Using Equation 8 $c_u = (0.6438)$mg/ml × $(5.60/6.35) = 0.568$ mg/ml. (Note that 3 significant figures are assumed for the depths so that c_u has only 3 significant figures. In practice the ruler can be read to the nearest 0.05 cm.)

The same calculation of c_u is to be repeated for the 6–9 trials. The values should lie within 10–15 percent of each other. The *average value* of c_u for 6–9 trials is reported along with the standard deviation and precision of the mean if these latter terms have been explained.

Report the concentration of the unknown Cu^{2+} solution in milligrams per milliliter. Calculate the standard deviations of your results.

IV. ERROR ANALYSIS

Calculate the maximum uncertainty in the copper analysis that arises from the following sources:

1. The error in weighing the copper wire for the standard solution is ±0.001 g.

2. An error of ±1.0 ml occurs in making up the volume of the standard solution.

3. An error of ±2 mm is made in measuring the height of the known solution using the visual method of estimating the point of equal intensities.

What would be the nature of the error if

4. a less concentrated solution of ammonia were used in the known compared to the unknown solutions?

5. the unknown solution were prepared one day, stored in an open flask, and used two days later, whereas the known solution was prepared and used on the day it was prepared?

NAME **DATE**

SECTION

A. Concentration of standard solution

g of Cu wire ______

c_k, concentration of Cu^{2+} (mg/ml) ______

B. Table of data

Trial	l_k Known solution depth (mm)	l_u Unknown solution depth (mm)	c_u Unknown solution conc. (eq. 8)
		average =	
		σ =	

NAME   **DATE**

SECTION

SELF-STUDY QUESTIONS

1. Sketch the structure of the copper complex ion used in this experiment.

2. Give a balanced equation for the reaction used to dissolve copper when preparing the known solution.

3. Suppose a standard solution were prepared by dissolving 0.2871 g metallic copper in 250 ml of solution using appropriate chemical reactions.
   a. How many moles of copper would be present in the solution?

b. What would the molar concentration of copper ion be in the solution?

4. Suppose the standard solution prepared in problem 3 were used to match the intensity of an unknown solution, the depth of the known and unknown solutions at equal intensities being 6.78 and 7.35 respectively.

a. What would the molar concentration of copper be in the unknown solution?

b. How many mg of copper would be present in each ml of unknown solution?

APPENDIX 7. THE BEER-LAMBERT LAW AND ITS USE IN QUANTITATIVE ANALYSIS

A. ABSORPTION OF RADIANT ENERGY BY MATTER

When a beam of light passes through a solution some of the energy may be absorbed by the molecules in the solution. If the beam emerging from the solution has less energy than when it entered, we say that some of the light has been *absorbed*. Visible light represents only a small part of the electromagnetic spectrum and consists of waves having wavelengths from 4000 Å to 8000 Å. All forms of electromagnetic energy occur in discrete bundles, called *photons* or *quanta;* the energy of a photon is proportional to the frequency of the radiation according to Equation 1, where

$$E = h\nu \tag{1}$$

E is the energy in ergs, ν the frequency of the radiation in cycles per second, and h is a universal constant—called Planck's constant—that has the value 6.624×10^{-27} erg-sec. As a matter of reference, ultraviolet radiation with short wavelengths (high frequencies) has a higher energy content than the long wavelength infrared radiation.

Molecules can interact with photons in three fundamental ways.

1. The energy of the photon is consumed to increase the rotational energy of the molecules.

2. The energy of the photon is used to increase both the vibrational and rotational energy of the molecule.

3. The energy of the photon is used to excite an electron (or electrons) in the molecule to higher energy levels; also the vibrational and rotational energy of the molecule may be affected.

The processes are listed in increasing order of energy—i.e., process 1 requires the least energy and process 3 the largest. Because of differences in their structures, molecules of different substances differ in the amount of energy required for each of the processes. Thus, photons of different energies (frequencies) are absorbed by some molecules but not by others, which leads to *absorption spectra* that are characteristic of each type of molecule. We are concerned here with the determination of the concentration of a molecule in a sample by the amount of light absorbed at some fixed wavelength (or frequency).

B. THE BEER-LAMBERT LAW

Consider a beam of *monochromatic* (single wavelength) light passing through a solution containing molecules which absorb the beam; the concentration of these molecules is C and the length of the beam path in solution is l. If the initial beam

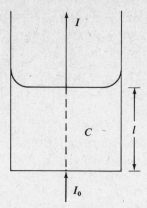

Fig. 7.1 *The important quantities which must be considered for the Beer-Lambert Law.*

intensity is I_0, a less intense beam I emerges. Figure 7.1 shows the relationship of these quantities.

The Beer-Lambert Law states that the intensity of the incident beam decreases exponentially with an increase in the thickness of the medium through which it passes and with an increase in the concentration of the absorbing molecules. Symbolically, the Beer-Lambert Law is given by Equation 2, using the quantities defined in Figure 7.1; ϵ is a constant, called the

$$\log \frac{I_0}{I} = \epsilon c l \tag{2}$$

molar extinction coefficient when the concentration c is expressed in terms of moles/liter. The term $\log I_0/I$ is called the absorbance A (or sometimes in the older literature the optical density, OD) of the solution. Thus, Equation 2 becomes

$$A = \epsilon c l \tag{3}$$

C. DETERMINATION OF CONCENTRATION USING THE BEER-LAMBERT LAW

Equation 3 can be used to estimate the concentration of an absorbing species in solution in several ways. Section E below gives a visual method based upon comparing the absorbance of an unknown solution with the absorbance of a solution with a known concentration of the species of interest. The absorbances of the two solutions are made equal by varying the length of the light path through these solutions. Under these conditions we can write Equation 4,

$$A_k = A_{uk} \tag{4}$$

where the subscripts k and uk stand for "known concentration" and "unknown concentration,'' respectively. Substituting Equation 3 into Equation 4 we obtain

$$\epsilon_k c_k l_k = \epsilon_{uk} c_{uk} l_{uk}. \tag{5}$$

Solving Equation 5 for c_{uk} (the unknown concentration) we obtain

$$c_{uk} = \frac{\epsilon_k}{\epsilon_{uk}} c_k \frac{l_k}{l_{uk}}. \tag{6}$$

But since we have the *same species* absorbing light in both tubes,

$$\epsilon_k = \epsilon_{uk} \tag{7}$$

and Equation 6 reduces to

$$c_{uk} = c_k \frac{l_k}{l_{uk}}. \tag{8}$$

Thus, the unknown concentration is equal to the known concentration multiplied by the appropriate ratio of light path lengths.

There are several kinds of instruments available to assist in the comparison of light intensities absorbed by solutions. Some—called *colorimeters*—and visual comparison using combinations of optical and mechanical devices. Others, called *spectrophotometers,* use electro-optical devices to make the comparison. You may have an opportunity to use both types in this course. If you are assigned to one of these instruments, the appropriate following directions should be followed.

☛ **CAUTION:** Handle these instruments carefully; they are delicate and will not suffer abuse.

D. COLORIMETER INSTRUCTIONS

The following instructions for the use of the *AO** Spencer Direct Result Colorimeter have been abstracted from the instruction manual.

Description: The essential parts and adjustments of the colorimeter are shown in Figure 7.2. Light from an incandescent bulb in the base of the instrument passes through illuminating windows under each plunger. The relative brightness of the two beams is adjustable by means of a light control knob. Once adjusted, the control knob may be locked in position by means of a clamping screw.

With the cups in place, light passes through the liquid held in the cup and enters the plunger. The depth of the liquid through which the light travels in the cup is adjustable by raising or lowering the cup. This is done by rotating the corresponding graduated drum on the side of the instrument.

The top of each plunger is beveled to form a prism which bends the beam of

---
*American Optical Company.

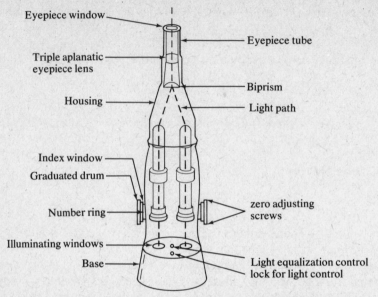

Fig. 7.2 *A diagrammatic representation of a colorimeter.*

light passing through it, so as to illuminate the corresponding side of the biprism. The dividing edge of the biprism forms the division between the illumination furnished by the beams of light passing through each plunger-cup system. The relative intensity of illumination on each side of the biprism is observed by means of the eyepiece, which contains a suitable filter for modifying the light to a quality approximating daylight.

Adjusting illumination: Before using the colorimeter for the first time, or after replacing the bulb, check the equality of illumination. The accuracy of readings made with a colorimeter is affected by the evenness of the illumination. The illumination can be adjusted approximately by removing the cups and adjusting the position of the lamp with equalization control knob, until the two sides of the field appear equally bright. To obtain higher precision, fill both cups with portions of the *same* colored liquid. Set one cup to a convenient depth, and take several readings of the position of the other. If the average of these differs from the depth of the other cup, the position of the lamp can be altered slightly until the average reading is the same as the reading on the fixed cup. This ensures that the illumination is set with an accuracy comparable to that produced by averaging several readings on the sample. A locking screw is provided to make the adjustment permanent.

NOTE: Only light clamping is required. In order to prevent casual or accidental interference with the light control, the locking screw is purposely made to require the use of a screw driver.

Filling the cups: The fused cups are one-piece construction, having a bottom plate fused onto the sides of the tube and the whole cup permanently cemented in a metal base. The cups fit into recesses in the cup carriages and can be inserted when the carriages are in their lowest position. If the cups are filled only to the top of the narrow cylindrical portion, they will contain enough liquid for most measurements and will be in no danger of spilling.

Reading the scale: The cup carriages are controlled by knobs having unit and decimal scales. The integer scale on the number ring moves continuously at a rate slower than the decimal scale on the drum. The drum scale is divided into several 100-part intervals with marks every fifth unit. A reading of cup position consists of a whole number reading, represented by an integer on the unit scale, followed by the decimal reading taken from the drum scale. The window carries an index mark against which both integral and decimal scales are read. In order to obtain correct reading, it is essential that the metal cup bottoms bear evenly in the carriage recesses. Any dirt or grit between these surfaces will cause zero errors. If necessary, the zero can be adjusted slightly by loosening the screws on the side of the knobs and moving the scales. The scale drums are more closely calibrated (¼ mm units) than the customary millimeter scale in order to simplify the calculation of the answer. If it should be necessary for any purpose, the scale reading can be converted to millimeters by multiplying these scale readings by four.

Calculation of results: In the customary use of any plunger type of colorimeter, one cup is filled with the unknown solution and the other is filled with a standard solution containing the same ingredients in a known concentration. If the standard and unknown are not too dissimilar in concentration, a suitable depth of solution can be found for each which will produce identical colors in the eyepiece of the colorimeter.

After the depths of the standard and unknown solutions have been adjusted to give matching fields in the eyepiece, the concentration of the unknown solution is calculated according to Equation 8. The scale readings are used in place of the light path readings, l_k and l_{uk}.

E. VISUAL (NON-INSTRUMENTAL) METHOD FOR COLORIMETRY

A relatively simple, non-instrumental technique may be applied when there is only one colored species present in solution. The method requires two viewing tubes, which may be test tubes, a source of light, and your eye (Figure 7.3); use the following procedure.

1. A colored solution (the unknown solution) is placed in a viewing tube and the color intensity compared to a solution containing the same colored species at a known concentration in an identical viewing tube. We assume in this discussion that the standard solution is more deeply colored than the unknown solution.

2. The two tubes are viewed down their lengths and solution is withdrawn from the standard solution until the color intensities are matched. The solution depth in

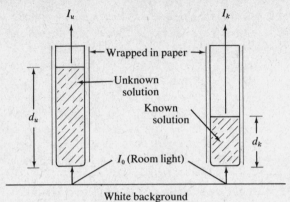

Fig. 7.3 *Visual method for colorimetry.*

tubes containing the known and unknown solutions is noted. The concentration of the unknown solution is calculated according to Equation 9.

$$c_{\text{unknown}} = \frac{c_{\text{known}}\ l_{\text{known}}}{l_{\text{unknown}}} \tag{9}$$

where c = concentration of the solution indicated
        l = solution depth at equal color intensity.

The greatest accuracy results when the solution depth is as great as possible while still allowing sufficient light transmission to detect the color. It is best to avoid lighting from the side (i.e., wrap test tubes in paper), and view against a white background (a piece of white paper will do). The greatest source of error is in deciding when two color intensities match. It is best to adjust the solution depth of the known solution around the point where it appears "obviously more intense" and "obviously less intense" than the unknown, before making the final "equal intensity" depth adjustment. One should repeat each measurement three times to ensure reasonable accuracy.

F. SPECTROPHOTOMETER INSTRUCTIONS

There are a variety of spectrophotometers and they all have the basic components shown in Figure 7.4. The source of monochromatic light can be as simple as a colored glassfilter inserted in front of a tungsten lamp; at the other extreme it could be a prism through which white light passes, with a mechanical device that allows only a specified wavelength to be "tuned in." These instruments are generally operated by first placing a sample of solution with a known concentration in the light beam to calibrate the read-out device, followed by the unknown sample. In some instances there is a sliding carriage to permit the ready interchange of sample and reference cell, whereas in other instruments the cells are manually interchanged. The following directions have been abstracted from the instruction manual for the Coleman® Junior Spectrophotometer, which you may use in this course.

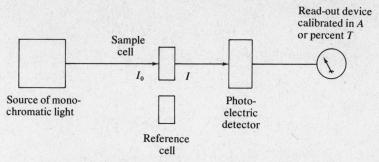

Fig. 7.4 *The components present in a typical spectrophotometer.*

TO MEASURE TRANSMITTANCE (*T*) OR ABSORBANCE (*A*) USING A COLEMAN® JUNIOR:

1. Insert the general purpose scale panel if it is not present. This will usually have been done by the time you get to the instrument.

2. Insert in the cuvette well a cuvette adapter of the proper size to accept the type of cuvette specified in the contemplated analytical method.

NOTE: A cuvette *is a specially constructed test tube. In our experiments we will use ordinary test tubes for our work. Be certain to select test tubes that are clear of scratches or faults.*

---

3. Turn on the instrument.

4. Verify the galvanometer zero setting and readjust if necessary.

5. Adjust the λ dial to the wavelength specified by the contemplated analytical method.

6. Carefully wipe the lower third of a cuvette containing at least the minimum volume of reference solution, which is usually the solvent, and properly position it in the cuvette well.

7. Adjust the "GALV Coarse" and "Fine knobs" until the galvanometer index reads the value for the reference specified in the contemplated analytical method. Usually this will be 100 percent T on the black Transmittance scale, or zero if the red absorbance scale is used.

8. Remove the cuvette of reference solution and replace it with a similar cuvette containing sample solution, wiped clean and properly positioned as before.

9. Read the *T* or *A* of the sample from the position of the galvanometer index on the same black or red galvanometer scale as was used for the initial adjustment (7).

Steps 6 through 9 should be done for both the known solution and the unknown solution—that is, you should determine values of *A* (or *T*) for both known and unknown solutions. These data can be treated in the following way to obtain the concentration of the unknown solution. From Equation 3 the absorbances of the known and unknown solutions are given by Equations 10 and 11, respectively.

$$A_k = \epsilon_k c_k l_k \tag{10}$$

$$A_{uk} = \epsilon_{uk} c_{uk} l_{uk} \tag{11}$$

Dividing Equation 10 by Equation 11 we obtain

$$\frac{A_k}{A_{uk}} = \frac{\epsilon_k}{\epsilon_{uk}} \frac{c_k}{c_{uk}} \frac{l_k}{l_{uk}}. \tag{12}$$

Since both known and unknown contain the same species

$$\epsilon_k = \epsilon_{uk}. \tag{13}$$

and Equation 12 becomes

$$\frac{A_k}{A_{uk}} = \frac{c_k}{c_{uk}} \frac{l_k}{l_{uk}}. \tag{14}$$

We recognize that the light path through the two cuvettes (test tubes) containing the known and unknown solutions are the same, if the same diameter tubes are used; thus

$$l_{uk} = l_k \tag{15}$$

and Equation 14 becomes

$$\frac{A_k}{A_{uk}} = \frac{c_k}{c_{uk}}. \tag{16}$$

Rearranging Equation 16 gives

$$c_{uk} = c_k \frac{A_{uk}}{A_k}. \tag{17}$$

Equation 17 says that the concentration of the unknown solution is equal to the known solution, multiplied by the ratio of the appropriate absorbances of the two solutions.

If percent T is measured, the same analysis obtains, if we recognize that the relationship between A and percent T is given by Equation 17

$$A = \log \frac{100}{\text{percent T}} \tag{18}$$

Thus we can calculate A from percent T using Equation 18 and then use these values of A with Equation 17.

II

GENERAL EXPERIMENTS

Atomic Structure

EXPERIMENT

Identification of Elements from Atomic Emission Spectra

I. INTRODUCTION

This experiment illustrates (1) the nature of the *line spectra* of atoms, one of the important experimental observations that lead to the *quantum theory of matter* and (2) the utility of this phenomenon in characterizing and identifying elements. There are several major features of this experiment: A *spectrograph** (Fig. 8.1) will be used to separate the light emitted by an atomic lamp into its various *wavelengths*. For atomic lamps, it will be found that only certain wavelengths will be emitted, resulting in a set of brightly colored lines being observed in the spectrograph (hence the name ''line spectra''). A lamp containing a known substance will be used to calibrate the position of the various emission lines on the spectrograph. The spectrum of a lamp containing an unknown substance will be obtained and the wavelengths of the various emission lines calculated. By comparing these wavelengths to the wavelengths of elements in Table 8.1, it will be possible to identify the element in the unknown lamp. The various components of the experiment are discussed in the following sections.

A. The Spectrograph

A spectrograph is an instrument designed to disperse or separate light according to its wavelengths. There are a variety of spectrographs available but they all rely

---

*The present discussion centers on the use of a focussing grating spectrograph. Other types of spectrographs, or even a monochromator, may be used. These are discussed in Appendix 8.

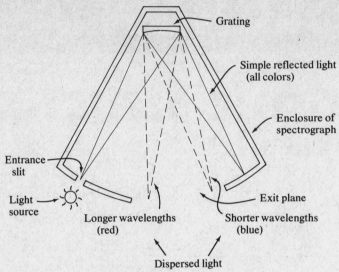

Fig. 8.1 *Spectrograph as viewed from above.*

Table 8.1
Principal Emission Wavelengths of Various Atomic Lamps
(Visible Region); S = strong, D = diffuse.

Element	Line position, Å				
Ag	4055	5209S	5467D		
Au	4793	5837	6278S		
Bi	4723S				
Ca	4227	4455			
Cd	4678	4800	5085S	6099	6438S
Ga	4033S	4172S			
Gd	4326				
Ge	4226				
H	6563	4861			
Hg	4358	5461			
He	5875S				
I	4862	5119	6082S	6566	
Ir	4102	4268			
Li	4132	4602	4971	6103S	6707S
Mg	5184S				
Na	5892S				
O	4368S	5331	5437	6158S	6456
Os	4261	4294	4420		
Pd	4087	4788	4213	5164	5296
Pt	4119	4443			
Rn	4307	4350S	7055		
Se	4736D	6303S	6326		
Ti	4981	5866	6146	6260	
Tl	5350S				
Xe	4624S	4671S	4734		
Zn	4680	4722	4811	5182	6362S

on the same basic principles. The basic parts common to all spectrographs are illustrated in Figure 8.1.

1. entrance slit—a narrow opening through which the light from the source to be studied is passed;

2. focussing grating—the most important part of the spectrograph is a curved mirror with a large number of vertical grooves cut on its surface (15,000 grooves/inch for the one employed here);

3. exit plane—the area of the spectrograph where the dispersed light is focussed. It is this area that will be observed by the student.

The operation of the spectrograph can best be explained by referring to Figure 8.1, which is a drawing of the spectrograph as viewed from above. A large portion of the light that strikes the grating is simply reflected, just like a normal mirror. However, the grooves on the surface of the grating cause a part of the light to be scattered in all directions away from the front face of grating. Owing to the wave nature of light, constructive or destructive interference of the scattered light occurs. Thus, for certain angles the intensity of light at a particular wavelength is low. Since the angle of constructive scattering is wavelength dependent, the scattered light is separated according to wavelength and the light is said to be *dispersed*.

This explanation is too brief to be helpful unless you have some knowledge of light, wave behavior and interference effects, and elementary optics. Usually these subjects are discussed in introductory physics courses. For the student not previously exposed to these subjects it is sufficient to realize that the spectrograph serves to spread out the light from the source across the exit plane according to its wavelength. It is assumed here that the student is familiar with the concept of wavelength as applied to light.

What is actually seen when the spectrograph is in operation is illustrated in Figure 8.2, which is a view of the side of the spectrograph that contains the entrance slit and exit plane.

The *position* of the various bright lines that appear at the exit plane are measured by placing a piece of ordinary clear or semi-transparent adhesive tape ("Scotch tape") across the exit plane and marking the position of the lines with a suitable pen. The tape is then removed and stuck onto a page in the laboratory notebook, care being taken to keep the tape smooth. The appearance of the piece of tape is given in Figure 8.3 for a hypothetical example. The vertical lines indicate

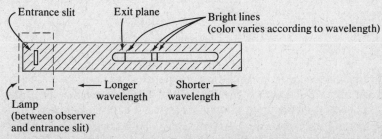

Fig. 8.2 *Spectrograph as viewed from the side.*

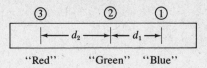

"Red"   "Green"   "Blue"

Fig. 8.3 *Tape marked with positions of emission lines.*

the positions of the emission lines indicated by "red," "green," and "blue"; d_1, d_2 are the distances measured in mm between the emission lines. The main useful property of the spectrograph for these experiments is that the position of the line as it appears in the exit plane is proportional to wavelength. The method of the calculation of the wavelength of an emission line is described in section III. The correspondence between wavelengths of light and the color seen by the human eye is listed in Table 8.2.

Table 8.2

$\lambda(\mathring{A})$	color
7500-6100	red
6100-5950	orange
5980-5800	yellow
5800-5000	green
5000-4350	blue
4350-3800	violet

B. The Atomic Emission Lamp

The lamp used in this experiment operates on much the same principle as a fluorescent lamp (which contains Hg) with the phosphor removed. A stream of electrons is accelerated through a low pressure of gas of the element of interest. In the case of metallic elements the initial action of the stream of electrons is to heat up the metal until a significant vapor pressure (perhaps a few mm) is present in the lamp. The electrons collide with the gaseous atoms and in some cases the electron-atom collision results in the atom being excited to a higher energy level. The excited atom may emit a photon, the frequency (λ) of which is determined by the relation given in Equation 1

$$\Delta E = \text{(energy excited state)} - \text{(energy lower state)} = h\nu \qquad (1)$$

where h is Planck's constant. From the theory of light we know that $\nu\lambda = c$ (λ = wavelength in cm, $c = 3 \times 10^{10}$ cm/sec, the speed of light) so

$$\Delta E = \frac{hc}{\lambda} \qquad (2)$$

These relations are often represented by an energy level diagram such as the type shown in Figure 8.4. The arrow connecting the two states is meant to depict the *transition* that occurs when light of wavelength $\lambda = hc/\Delta E$ is emitted. The energy level diagram and full explanation of the emission spectra for the elements to be studied here (except H, see next section) are far too complex to be included in an

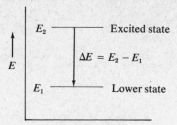

Fig. 8.4 *Energy level diagram.*

introductory course. Suffice it to state that the emission spectrum (and energy level diagram) of a particular element is unique to that element and serves to identify it.

C. The Hydrogen Lamp and the Balmer Series

One exception to the complexity mentioned in the last paragraph of the previous section is the hydrogen atom. The order of energy levels of the hydrogen atom are given by the Rydberg formula (Eq. 3)

$$E_n = -\frac{R}{n^2} \qquad n = 1, 2, 3, \ldots \tag{3}$$

where R is the Rydberg constant (in energy units $R = 2.18 \times 10^{-11}$ ergs/atom). This formula was originally derived empirically and was one of the starting points in the development of the quantum theory.

The predominate species in a hydrogen lamp is molecular hydrogen, H_2. However, the stream of accelerated electrons serves to dissociate the H_2 molecule; the hydrogen atoms formed in this process are then excited by the electron stream. The line spectrum observed arises from transitions from states with $n > 2$ to the $n = 2$ state. This series of lines is known as the *Balmer series,* after its discoverer. The series of transitions that terminate in the $n = 1$ level (the Lyman series) are at higher energy (shorter wavelengths) and are not visible to the naked eye. The transitions that are observed are depicted in Figure 8.5. The wavelength of the transitions of the Balmer series are indicated in parenthesis. The colors of these lines are (from left to right) red, blue-green, blue, violet.

D. Calibration of the Spectrometer

For the present experiment either the Balmer series from the hydrogen lamp or the light from a fluorescent lamp will be used to calibrate the spectrograph, which in turn will be used to establish the spectrum, and identity, of the unknown element. The red 6563 Å and blue-green 4861 Å in the hydrogen spectrum are by far the brightest hydrogen emission lines observed and should be used as calibration lines.

Fluorescent lamps contain mercury, which also emits a characteristic line spectrum. Most of this radiation is in the ultra-violet region and therefore is not visible to the human eye. Fluorescent lamps have a phosphor coating on their in-

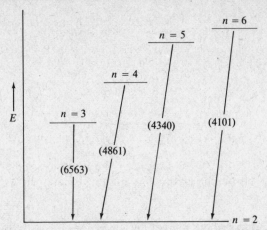

Fig. 8.5 *Balmer series. The wavelengths of the transitions of the Balmer series are given in parentheses in Å. The colors of these lines are (from left to right) red, blue-green, blue, violet.*

ner walls, however, which converts the ultra-violet radiation to visible white light. When a fluorescent lamp is used with the spectrograph, the white light is dispersed and appears at the exit slit as a weak continuous spectrum, shading from red to violet. However, two of the strong visible lines of the mercury emission spectrum "leak" through the phosphor and appear clearly as bright lines superimposed on the continuous spectrum. The wavelengths of these two lines are *5460 Å* (green) and *4360 Å* (violet) and also can be used for calibration.

II. PROCEDURE

The procedure to be followed is quite straightforward. It is assumed that the proper handling of the lamp and equipment has been demonstrated in lecture or by the laboratory assistant.

☛ CAUTION: DO NOT UNDERTAKE THIS EXPERIMENT UNLESS YOU HAVE WITNESSED THIS DEMONSTRATION. Be sure and note the date and time of the observation.

---

1. Place a piece of clear adhesive tape across the exit plane.
2. Position the known calibrating lamp (hydrogen lamp or a fluorescent lamp) next to the entrance slit of the spectrograph. If you use a hydrogen lamp, check the exit plane for the two brightest Balmer lines (the red line will be on the observer's left). A similar check should be made if the fluorescent lamp is used. Using a suitable pen, mark the position of the two brightest lines on the tape.
3. Position the lamp containing the unknown element next to the entrance slit. Mark the position of each emission line. Either use a different colored ink or suitable identification marks so that the lines from the unknown will not be confused with the lines from the calibration lamp. RECORD THE POSITION OF THE

BRIGHTEST LINES ONLY. Ignore the faint lines that can be perceived only with difficulty.

4. *Carefully* peel the adhesive tape from the exit plane and stick the tape on a page in your notebook, taking care to keep the tape flat.

NOTE: Be sure to note the date and time of the observation.

---

5. Calculate the wavelengths of the emission lines of the unknown lamp (see section III). Compare the measured wavelengths with those in Table 1 and on the basis of this comparison identify the element present in the atomic emission lamp (your wavelengths should agree within ±30 Å of those tabulated).

6. Take measurements on different days* and identify each unknown lamp. The data and time of day (AM or PM) of each observation must be noted on the report because the atomic lamps will be changed regularly and a record kept of the atomic lamp used on any given date.

III. DATA ANALYSIS

In the present setup for the atomic emission experiment only the red (6564 Å) and blue-green (4861 Å) hydrogen lines can be easily seen. The following example calculation shows how to calculate the wavelength of a single unknown emission line using the hydrogen spectrum as a reference; a similar argument obtains for experiments using a fluorescent lamp (Hg lines) as the standard. Most emission lamps will have more than one emission line, but the calculation for each individual line follows the same pattern.

1. The drawing given as Figure 8.6 is a representation of an experimental spectrum. All distances are in mm. The accuracy of measurements is about ±1 mm. Note that 41 + 68 = 109 mm which is within 1 mm of 108 mm.

2. The dispersion of the spectrograph is calculated as follows:

$$\frac{\text{wavelength difference of two hydrogen lines}}{\text{distance in mm between two hydrogen lines}} = \frac{(6563 - 4861)}{108} \frac{\text{Å}}{\text{mm}}$$

$$= \frac{15.8 \text{ Å}}{\text{mm}}.$$

---

*Exact instructions will be given by your laboratory instructor.

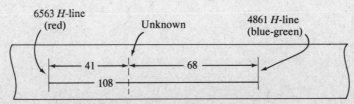

Fig. 8.6 *Representation of an experimental spectrum.*

The value of the dispersion will depend upon the spectrograph used.

3. By inspection we see that the wavelength of the unknown line is shorter than 6563 Å and longer than 4861 Å. The unknown wavelength can be calculated two ways:

$$\lambda = 6563 \text{ Å} - (15.8 \text{ Å/mm}) (41 \text{ mm}) = 5915 \text{ Å (relative to 6563 Å line.)}$$
$$\lambda = 4861 \text{ Å} + (15.8 \text{ Å/mm}) (68 \text{ mm}) = 5935 \text{ Å (relative to 4861 Å line.)}$$

NOTE: Be sure to note that different signs are used in the calculations.

The agreement between these two calculations is 20 Å, which is about the level of accuracy of the technique. The average of these two values (Eq. 1) should be taken for the final value of the wavelength:

$$\lambda_{ave} = \frac{5915 + 5935}{2} = 5925 \text{ Å.}$$

4. When you compare the wavelengths you obtain with those listed in the table in Appendix 10, don't forget your accuracy is approximately 20 Å, so any value within ±20 Å of a listed value constitutes a "match."

5. In recording data don't forget that only the strong emission lines should be noted. You should never see more than 4 or 5 "strong" lines. Weak lines should be disregarded (some people with less sensitive eyes won't see them at all).

IV. ERROR ANALYSIS

Two types of errors are possible in this experiment: (1) gross errors, where a line is not seen because it is relatively weak and (2) measurement errors in the separation of the emission lines. In the following, estimate the effect of various measurement errors on the calculated wavelength of an emission line.

1. Suppose there is an error of ±2 mm on the separation of the lines used to calibrate the spectrograph. What effect will that have on the calculated dispersion of the spectrograph?

2. Suppose there is an error in the measured position of an unknown emission line by ± 1 mm. What will be the resulting error in the calculated wavelength of the unknown emission line? You will have to use your measured value of the dispersion to estimate this error.

3. What is the maximum error that can result from a combination of 1 and 2?

4. If it is allowed by your laboratory instructor, move the lamps with respect to the entrance slit by a small amount. Is there any visible movement of the position of the emission line? If so, estimate the maximum uncertainty in your data from variable placement of the atomic emission lamps.

NAME   **DATE**

SECTION

SELF-STUDY QUESTIONS

1. Sketch a spectrograph and discuss how the spectrograph separates light according to wavelength.

2. What is meant by the term *line spectrum*? Explain how the observed line spectrum from a lamp may serve to identify the element(s) in the lamp.

3. By means of a general energy level diagram, explain the types of transitions involved in the emission of light.

4. What is meant by the term *dispersion of the spectrograph*? What are the units of this quantity?

5. Suppose you knew the distance between two emission lines on the exit plane of the spectrograph. What could be calculated about these two lines if you knew only the dispersion of the spectrograph? What further information is required to obtain the wavelength of the two emission lines?

6. Explain how the observation of the hydrogen 6563 Å and 4861 Å emission lines, or the mercury 5460 Å and 4360 Å lines can be used to calibrate the spectrograph.

APPENDIX 8. OTHER SPECTROGRAPH DESIGNS

A. Bunsen Spectroscope

The Bunsen spectroscope is illustrated schematically in Figure 8.7. The dispersive element for this device is a prism (rather than a grating, as is the case with the spectrograph described in the main text). The major feature of the spectroscope is that one may vary the angle between the observer to prism (line *O–P*) and the source to prism line (line *S–P*). The angle at which the light is refracted by the prism will depend on the wavelength, and hence at certain angles a line (or lines) corresponding to the emission lines of the element will be observed. One important difference between a prism and a grating device is that the dispersion of the former is not linear, i.e., the angular separation between two emission lines depends not only on the wavelength difference of the lines but on their wavelengths. For that reason a much more elaborate calibration of the instrument is required. It is suggested that an He emission tube be used, and the wavelengths, colors, and relative intensities of the He emission tube are given in Table 8.3. As many of these lines as possible should be observed, but the weaker ones may be difficult unless the room is darkened and the eye is protected from the full output of the He lamp.

In obtaining the wavelength of the unknown element, the following is suggested:

1. Prepare a calibration curve of wavelength vs. angular position from the data obtained from the He lamp.

2. Using the calibration curve and the observed angles of the emission lines for the unknown element, estimate the emission wavelengths for the unknown, and hence identify the unknown.

B. Homemade, Transmission-Grating Spectrograph

Figure 8.8 is based on a 35 mm photographic replica of a transmission grating, mounted in a standard 2 in. × 2 in. slide holder.* Dimensions shown are approximate, and experimentation with different configurations is certainly possible.

---

*Such gratings have been available from science supply houses such as Edmund Scientific.

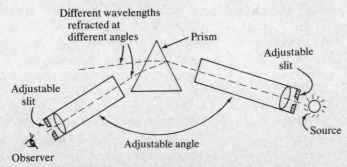

Fig. 8.7 *Schematic of Bunsen Spectroscope.*

Table 8.3
He Discharge Lamp Lines*

Wavelength (A)	Color	Relative Intensity**
7281	red	30
7065	red	70
6678	red	100
5876	yellow	1000
5048	green	15
5016	green	100
4922	green	50
4713	blue-green	40
4471	blue-violet	100
4438	blue-violet	10
4388	violet	30

*Many of these lines actually arise from molecular He species (such as He_2^+) formed in the discharge tube.

**Relative intensities are highly qualitative and depend on the conditions used for the discharge lamp and the spectral sensitivity of the observer's eye.

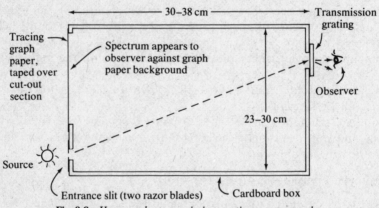

Fig. 8.8 *Homemade, transmission-grating spectrograph.*

In use the entrance slit is held toward the source, and the observer looks straight through the grating (*not* at an angle that allows direct viewing of the source). The image of the lines of different wavelengths appears on the graph paper background. The grid of the graph paper serves as a scale to allow the measurement of the position of the lines. The calibration of this spectrograph follows the same procedure as the main text.

SECTION B

Synthesis

EXPERIMENT

9

Synthesis of
trans-Dichlorobis
(ethylenediamine)
cobalt (III) Chloride

I. DISCUSSION

In the present experiment we will carry out the synthesis of a cobalt complex, *trans*-dichlorobis (ethylenediamine) cobalt (III) chloride, the structural formula of which appears as:

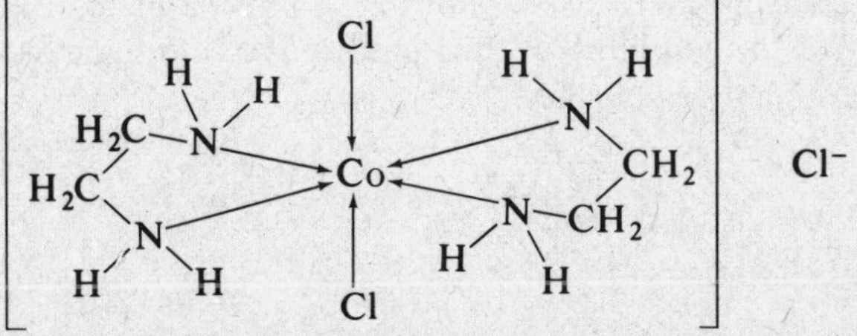

(I) (green)

In this ion, which is called an *octahedral complex* because the Co atom is surrounded by atoms situated at the corners of an octahedron, the nitrogen atoms are in the same plane as, and equidistant from, the Co atom. The groups attached to the cobalt ion, i.e., Cl^- and ethylenediamine, are called *ligands*.

In aqueous solution one of the chloride ions is rather easily

$$trans\text{-}[Co(NH_2CH_2CH_2NH_2)_2Cl_2]^+ + H_2O \longrightarrow$$
$$trans\text{-}[Co(NH_2CH_2CH_2NH_2)_2ClH_2O]^{2+} + Cl^- \quad (1)$$

replaced by a water molecule, forming the pink ion shown as:

(II) (pink)

The synthesis of the *trans*-[Co(en)$_2$Cl$_2$]$^+$ complex will involve (a) oxidizing the Co(II) complex [Co(en)$_2$(H$_2$O)$_2$]$^{3+}$, (b) addition of an excess of HCl, (c) removal of most of the water from the reaction mixture, forming the green dichloro complex ion II, and (d) the purification of the green crystals of *trans*-[Co(en)$_2$Cl$_2$]Cl by recrystallization.

A number of comments need to be made concerning this synthesis. The Co(II) ion does not bind ligands (i.e., the Cl$^-$ or ethylenediamine groups) as strongly as the Co(III) ion. The oxidation is carried out on the ion [Co(en)$_2$(H$_2$O)$_2$]$^{2+}$ rather than aqueous Co(II) (which is actually Co(H$_2$O)$_6^{2+}$ in solution), because the ethylenediamine ligands stabilize the Co(III) ion, which otherwise would react with water to produce O$_2$ according to Equation 2.

$$4Co^{3+} + 2H_2O = 4Co^{2+} + O_2(gas) + 4H^+ \quad (2)$$

The recrystallization purification step (d), is a standard procedure in chemical synthesis and is based on the principle that a pure crystal is more stable than one containing impurities. Thus, when an impure solid is dissolved in a suitable solvent and crystals are allowed to reform, the impurities tend to remain in solution (assuming that the solubility of the impurity is of the same order of magnitude as the substance sought in a purified state).

II. PROCEDURE

1. Using a triple beam balance, weigh out ~12 g of CoCl$_2 \cdot$6H$_2$O, place in a 250 ml beaker, and dissolve in about 25 ml of H$_2$O with stirring.

2. Add ~40 ml of 10 percent ethylenediamine solution. The solution will be observed to darken considerably as the [Co(en)$_2$(H$_2$O)$_2$]$^{2+}$ complex is formed. After ~10 minutes, cool the solution in an ice bath for 15–20 minutes before proceeding to step 3.

3. Add 10 ml of *30 percent* H_2O_2, a few ml at a time, to the cold cobaltous solution from step 2 while stirring the reaction mixture gently. Leave the cobaltous solution in the ice bath during the addition of H_2O_2. Bubbles of O_2 will be observed as H_2O_2 disproportionates according to Equation 3.

$$2H_2O_2 = 2H_2O + O_2 \tag{3}$$

Hydrogen peroxide, H_2O_2, reacts with the Co(II) ion according to Equation 4; in this equation

$$2Co(II) + H_2O_2 + 2H^+ = Co(III) + 2H_2O \tag{4}$$

it is assumed that the metal ions are suitably complexed although this fact is not explicitly shown.

After all the H_2O_2 has been added, remove the reaction beaker from the ice bath and allow the solution to stand for ~10 minutes. At this stage the solution is a disagreeable brownish color.

☛ **CAUTION:** 30 percent H_2O_2 should not be allowed to come into contact with skin or clothes. If it does, flush the area of contact with copious amounts of water immediately.

---

4. Slowly add 40 ml of concentrated HCl (12 *M*) to the reaction mixture, which will probably cause further bubbling as more H_2O_2 reacts to form O_2 according to Equation 3. Allow the mixture to stand 20–30 minutes; the reaction mixture can be stored at this point.

5. It is the object of this step to evaporate the reaction mixture to about one-third of the original volume. This step essentially forces the reaction, shown as Equation 5, to form product:

$$[Co(en)_2(H_2O)_2]^{3+} + 2Cl^- \longrightarrow trans\text{-}[Co(en)_2Cl_2] + 2H_2O \tag{5}$$

It is a relatively slow step, however, and requires some care. Transfer the reaction mixture to a 4 in. evaporating dish, and set up the arrangement sketched in Figure 9.1.

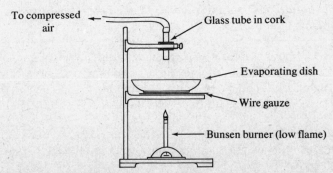

Fig. 9.1 *Evaporation of the mixture.*

NOTE: Fumes of HCl will be given off during the evaporation step. If at all possible carry out the evaporation under the hood.

For a successful evaporation the following precautions should be observed:

 a. Maintain a low flame and never allow the solution to boil violently (avoid splattering).

 b. Maintain a gentle flow rate of air and *always* turn the air on with the air flow directed *away* from the liquid until the air flow rate has been adjusted satisfactorily. At the end of the evaporation step the solution will be a dark purple color. It will then turn dark green as the chloride ions replace the water molecules in the complex according to Equation 5. The reaction mixture may be stored at any stage of this step.

 6. The hot solution may be transferred to a 250 ml beaker (use tongs) and chilled in an ice bath. After the solution is cool (below room temperature), vacuum filter the reaction mixture and wash the solid with ~20 ml of concentrated HCl. Save the solid material. The solid should be predominantly green but may be contaminated with areas of blue or purple, which arise from a variety of other Co(III) or Co(II) complexes. Remove the filter paper from the Buchner funnel and scrape the solid onto a watch glass. Weigh the moist material and record this in your notebook. The solid may be stored at this point. Use a triple beam balance.

☛ **CAUTION:** Do not carry out a vacuum filtration unless this procedure has been demonstrated in laboratory or lecture.

---

 7. The object of this step is the purification of the product by recrystallization. Make up a solution of 30 ml of conc. HCl and 10 ml H_2O. Transfer about 3–4 ml of this solution per gram of solid obtained from step 6 to a 150 ml beaker and bring the mixture to a boil; the solid product may require considerable stirring to help it dissolve in this hot solution. After 3–4 minutes stirring decant the hot solution into a clean 250 ml beaker and chill the decantate in an ice bath. Save any solid material left in the beaker after decantation. Allow the decanted solution to remain in the ice bath for 30 minutes before carrying out step 8.

NOTE: Your laboratory instructor will give detailed directions on the technique of vacuum filtration.

 8. There should be a crop of bright green crystals in the chilled solution from step 7. Filter this solution with the aid of vacuum and save the filtrate. If undissolved material is available from step 7, combine the filtrate and the solid in a clean beaker. The recrystallized product should be washed in the Buchner funnel as follows:

☛ **CAUTION:** Remove the filtrate from the vacuum flask before carrying out these steps.

a. With the vacuum off, pour ~10 ml of concentrated HCl over the crystals while they are still in the Buchner funnel and allow the mixture to stand for ~30 sec before the vacuum is restarted and the HCl solution is pulled through the Buchner funnel.

b. Break the vacuum and carry out two successive washes with 95 percent ethanol, following the same procedure as described for the previous wash (a). After the second washing with ethanol leave the vacuum on for 5–10 min to dry the crystals.

9. Scrape the green crystals onto a watch glass, allow them to air dry for ~30 minutes, and weigh them on the triple beam balance. Record the weight in your notebook. At least 0.5 g of product should be obtained. Turn in the product to the laboratory instructor.

10. A second crop of crystals can be obtained from the remaining filtrate by reducing its volume and/or dissolving any undissolved product from step 7, and repeating the recrystallization process. If you obtained more than 3 g of product at step 9, this procedure is not necessary. If you do obtain a second crop of crystals, keep the first and second crop *separate* because the first crop will be much purer than the second. You may be asked to save the first crop of crystals for a subsequent experiment.

III. DATA ANALYSIS

The yield (or percent-yield) is usually estimated for any experiment in synthesis. The theoretical yield is what might be expected if all of the starting material (in this case $CoCl_2 \cdot 6H_2O$) were converted to the desired product $[Co(NH_2CH_2CH_2NH_2)_2Cl_3]$. If the overall process were 100 percent successful, one mole of the starting Co-containing compound would yield one mole of the *trans* complex (Eq. 6).

$$CoCl_2 \cdot 6H_2O \longrightarrow Co(NH_2CH_2CH_2NH_2)Cl_3 \qquad (6)$$

Starting the synthesis with 12 g of $CoCl_2 \cdot 6H_2O$, which corresponds to 12/ (mw of $CoCl_2 \cdot 6H_2O$) moles, you would expect to obtain the *same* number of moles of $Co(NH_2CH_2CH_2NH_2)Cl_3$ if the whole process, *viz.*, the oxidation step (Eq. 4), the conversion of the aqueo complex to the chloro complex (Eq. 5), and isolation of the product from solution were to go to completion. However, we would expect that some of these intermediate steps (such as the isolation of the complex from solution) do not occur completely. Therefore the actual yield will be less than expected if all steps were perfectly accomplished. A measure of the efficiency of the synthesis is the percent yield which is given by

$$\text{percent yield} = \frac{\text{actual yield}}{\text{theoretical yield}} \times 100 \qquad (7)$$

Calculate the percent yield for your preparation of *trans*-$Co(NH_2CH_2CH_2NH_2)_2Cl_3$.

IV. ERROR ANALYSIS

Error analysis for synthesis experiments generally has a different character than that for physical measurements. Physical measurements often attempt to estimate errors in the measured quantities and to trace these errors as they affect the final result. The following questions have to do with factors that affect the yield of the synthesis.

1. Which of the steps in the preparation would you expect to lose the most product?

2. If the product is purple colored would you expect the yield to be large? Explain.

3. Explain what you would expect to happen to the yield under the following conditions.
   a. 10 percent H_2O_2 was used.
   b. H_2O_2 was added to the reaction mixture before the ethylenediamine.
   c. The impure product is recrystallized from H_2O.
   d. The recrystallized product is washed with H_2O.

NAME   DATE

_____

SECTION

_____

SELF-STUDY QUESTIONS

1. What is the oxidation state of the cobalt atom in the complex $[Co(en)_2 Cl_2]Cl$?

2. Write the half reaction for hydrogen peroxide acting as an oxidizing agent.

3. Explain which step in the preparation of $[Co(en)_2 Cl_2]Cl$ you would expect to be the most inefficient in the overall process.

4. Describe how you would destroy an excess of H_2O_2 in an aqueous solution. What would be the product?

5. What is the maximum yield of $[Co(en)_2Cl_2]Cl$ you could expect from 3 g of $CoCl_2 \cdot 6H_2O$?

6. Why must the product be recrystallized from an aqueous solution of HCl rather than H_2O?

7. The nomenclature of octahedral complexes can include the prefix *cis*- as well as *trans*-. *Trans*- signifies "opposite each other" whereas *cis*- signifies "next to each other." Give the structural formula for *cis*-dichlorobis (ethylenediamine) cobalt (III) chloride.

EXPERIMENT

10

The Synthesis of a Nitrite-Containing Coordination Compound

I. INTRODUCTION

This experiment involves the analysis of one of a group of compounds with the general formula $K_2MM'(NO_2)_6$, where M is a divalent nontransition metal ion and M' is a divalent transition metal ion. These compounds are of interest because they are examples of a very large class of compounds called *coordination*, or *complex, compounds*.

Complex ions are formed by the association of a metal ion with one or more molecular or ionic species known as *ligands* or *complexing agents*. The bonding in complex ions varies from predominantly ionic to predominantly covalent, depending upon the nature of the metal ion and the ligand involved. The transition metal ions show a strong tendency to form complexes with both negative ions and molecules as ligands, which is generally attributed to the relatively small size and large charge of the metal ions and to the presence of partially filled d orbitals in the valence shell which may be used to participate in covalent bonding. The number of bonding positions around the central metal ion which are occupied by ligand atoms is known as the *coordination number of the metal ion*. The coordination number of M' in $[M'(NO_2)_6]^{4-}$ is six with the NO_2^- ions arranged so that the six nitrogen atoms of the ligand surround the metal ion in an octahedral arrangement.

$$O_2N \diagdown \overset{\textstyle NO_2}{\underset{\textstyle NO_2}{\overset{|}{\underset{|}{M'}}}} \diagdown NO_2$$
$$O_2N \diagup \qquad \diagup NO_2$$

The bonds formed between the metal ion and the ligand are called *coordinate covalent bonds* because the ligand supplies both electrons to the bond formed between it and the metal atom.

Table 10.1

MCl$_2$ \ M'Cl$_2$	CaCl$_2$ (0.6 g)	SrCl$_2$ (0.8 g)	BaCl$_2$ (1.0 g)
NiCl$_2$ (0.7 g)	X	X	X
CoCl$_2$ (0.7 g)		X	X
CuCl$_2$ (0.7 g)			X

II. PROCEDURE

Synthesis of K$_2$MM'(NO$_2$)$_6$

Complex compounds of the type we are interested in here form readily with alkaline earth metals and certain transition metal ions. Successful combinations are indicated by the symbol X in Table 10.1. Your laboratory instructor will indicate which pair of metal salts you are to use in your preparation.

From the storeroom obtain samples of the transition metal and nontransition metal chloride assigned by your instructor. Weigh out the amounts indicated in Table 10.1. Combine the solids and dissolve the mixture in a minimum volume (5 ml) of distilled water. Gentle heating may be used *only* if needed to dissolve the salts. *After this operation you should have one solution containing two compounds.* Dissolve 4.25 g of KNO$_2$ in 5 ml of distilled water and slowly add this solution—with stirring—to the solution of metal chlorides. Cool the mixture in an ice bath. The product will be filtered using a vacuum filtering assembly (Fig. 10.1). Allow the solid formed to settle, then decant the supernatant liquid into a filter funnel, leaving most of the solid in the original container, where it can be more efficiently washed. Add to the solid a 3 ml portion of cold water containing about 0.1 g of KNO$_2$, stir well, and pour the mixture into the filter to collect the product.

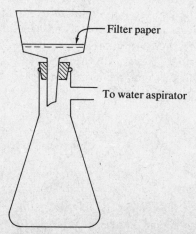

Filter paper

To water aspirator

Fig. 10.1 *Vacuum filtering assembly.*

When the water has all drained out of the filter, pour 5 ml of acetone over the solid to help it dry. After the acetone has drained through the filter, pour a second 5 ml portion of acetone over the solid. Complete the drying of your compound by using an aspirator to draw a stream of air through the filter gently. If no aspirator is available, open the filter paper and spread out the solid to dry in the air. The product is powdery when dry and has no odor of acetone. After drying the solid, determine its weight. The product should be dried over night in your desiccator before you attempt its analysis.

III. DATA ANALYSIS

Write a balanced equation for the formation of the nitrite complex compound using the salts assigned to you. Calculate the yield for your synthesis using the limiting (e.g., MCl_2, $M'Cl_2$, n KNO_2) reagent in the mixture.

IV. ERROR ANALYSIS

What would you expect would happen to the yield of the nitrite complex under the following conditions:

  1. the mixture of MCl_2 and $M'Cl_2$ is dissolved in 10 ml of water rather than 5 ml of water;

  2. the nitrite complex is precipitated by adding 20 g KNO_2;

3. the reaction mixture is cooled to room temperature before it is filtered;

4. the solid product is washed with H_2O;

5. the solid product is not washed with acetone, and is weighed immediately?

NAME   DATE

SECTION

SELF-STUDY QUESTIONS

1. Give a three-dimensional sketch of a hexanitrito-metal complex ion.

2. Write a balanced equation for the formation of the potassium salt of the hexanitrito complex you would expect to be able to form from KNO_2, $CuCl_2$, and $SrCl_2$.

3. What would you expect to happen if a student used KNO_3 instead of KNO_2 in the synthesis of this compound?

4. What would be expected to happen to his yield if a student washed his product with a cold solution of KNO_3?

Qualitative Analysis

11

Qualitative Analysis by Paper Chromatography

I. INTRODUCTION

Physical methods for identification depend upon differences in physical properties of the species in question. Often the differences in properties are very slight, and they are magnified by repeating the process many times. In this part of the experiment we shall use chromatography to separate and identify ions in an aqueous solution. *Chromatography* is the name applied to a large number of processes in which the components of a mixture are distributed between a stationary phase and a mobile phase. Most commonly, the stationary phase is solid and the mobile phase is liquid, although other combinations such as gaseous mobile phases and solid or liquid stationary phases, as in gas chromatography, are possible.

The adsorption (note it's *not* absorption) of foreign molecules on a surface occurs because of the existence of more or less strong forces between the molecules in the surface and the foreign molecules; these forces are similar in character to those which operate between the molecules of a liquid or those which govern the solubility of one substance in another. For the most part the relative polarity and polarizability of the interacting molecules are measures of the magnitude of the interactions. The intensity of adsorption of molecules on a surface depends upon the character of the surface. Thus, a properly chosen surface may selectively adsorb one component of a mixture rather than another. The law which governs the adsorption of a solute (which we may call A) from dilute solution is given by Equation 1

$$K_A = \frac{[\text{amount of solute } A \text{ adsorbed per unit surface area}]}{[\text{Concentration of } A \text{ in Solution}]} \qquad (1)$$

K_A is called the *adsorption coefficient.*

Consider a mixture consisting of two solutes dissolved in a solvent. This mixture contains three different kinds of molecules, and we would expect each molecule to exhibit different adsorption coefficients. (Only by accident would several molecules have the same adsorption coefficient.) Thus in this mixture the three kinds of molecules compete for adsorption sites on the solid phase with varying degrees of success. Suppose we could arrange a way to have this mixture move across a solid phase. In this case, the three kinds of molecules would initially adsorb on the surface according to their natural tendency to interact with the available surface sites and roughly in proportion to their concentration (Eq. 1). As the mixture moves across the surface, the initially adsorbed solute molecules would be constantly displaced by solvent molecules because the latter are present in a very large excess. There would, however, be a difference between the displacement of the two solute molecules because they have different values for K_A. Thus, if the mixture of solute molecules started at the same place on the solid phase, after a short time the two solute molecules would be adsorbed at different places; in other words, a separation would have occurred. This separation would be obvious if the two species were of different color or could give reactions which would produce different colors.

II. PROCEDURE

A very useful and simple method of chromatography, incorporating the ideas described above, is paper chromatography, which employs filter paper as the stationary phase and an appropriate solvent (or solvent mixture) as the liquid phase, which is called the *eluting solution.* In this part of the experiment we shall use paper chromatography to separate and identify the ions in a mixture of cations. Your unknown will contain one or more of the following cations: Fe^{3+}, Co^{2+}, Cu^{2+}, Mn^{2+}, Ni^{2+}.

A. Chromatographic Chamber

Assemble the apparatus shown in Figure 11.1 using a 19×150 mm test tube. Straighten a paper clip and push it slowly (**CAUTION**) but firmly through a cork

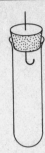

Fig. 11.1   *Apparatus using a 19 × 150 mm test tube.*

which fits the test tube. The bent paper clip hook will be used to suspend a piece of chromatographic paper in the test tube so that it just touches the eluting solution.

B. Preparation of a Micropipette

The unknown solution will be placed upon the chromatographic paper using a micropipette prepared by drawing a piece of glass tubing to a fine diameter.

NOTE: Your laboratory instructor will instruct you in the proper way to draw a piece of glass tubing using a Bunsen burner.

The micropipette should look like the object shown in Figure 11.2(B). Melting point capillaries can be used as micropipettes if they are available. The drawn end should be sufficiently small in diameter to draw liquid into it spontaneously when it is placed in a solution; no drops of liquid should form at the tip when it is removed from solution.

C. Preparation of Eluting Solution

Obtain 4.5 ml of acetone in a graduated test tube and add sufficient 6N HCl to bring the total volume of solution to 5 ml. Pour enough of this solution into a 19 × 150 mm test tube so that the level of solution is about 1 cm from the bottom; stopper this test tube and save it.

D. Preparation of the Chromatogram

From the stockroom, obtain a sample of the unknown in a large, clean, dry test tube, and a strip of chromatographic paper.

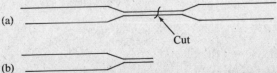

Fig. 11.2   *Micropipette can be prepared by drawing a piece of glass tubing to form a long constriction. Cutting at the point indicated produces two micropipettes similar to the one in B.*

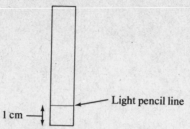

Fig. 11.3 *A piece of chromatographic paper before the unknown is spotted onto it. The unknown is applied along a lightly penciled line about 1 cm from the bottom of the strip.*

☛ **CAUTION:** Do not handle the paper with your fingers if either they or the paper, or both are wet.

With a pencil, draw a light line on one end of the strip of chromatographic paper about 1 cm from, and parallel to, the short dimension (Fig. 11.3). Using your micropipette, place a thin line of unknown solution along the pencil line on the chromatographic paper. When you place the end of your micropipette into the unknown solution, a column of liquid should be drawn up into the pipette. When you touch the tip of the pipette to the dry chromatographic paper, the liquid will be absorbed by the paper and be drawn from the pipette. You should make a few practice trials drawing a straight line with your unknown on ordinary filter paper.

E. Elution of the Chromatogram

After the chromatogram has dried for a few minutes, catch the end (opposite the unknown end) of the strip of chromatographic paper over the paper clip hook prepared earlier; the paper should hang freely at such a height as to dip into the eluting solution about 0.5 cm when the cork is placed in the test tube containing the previously prepared eluting solution. The apparatus should look like that shown in Figure 11.4.

Set the test tube in a vertical position for about 10 minutes or until the line marking the solvent front is near the top of the chromatographic paper strip.

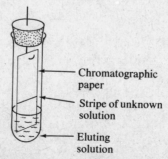

Fig. 11.4 *Eluting the chromatogram.*

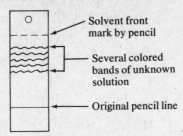

Fig. 11.5 *The eluted chromatogram will have several more or less distinct stripes of colors across it.*

F. Preparation of the Development Chamber

While the chromatogram is eluting, place about 0.5 ml of 6 N NH$_3$ in a 19 × 150 mm test tube and stopper it. Save for future use.

G. Marking the Chromatogram

When the solvent front has reached almost to the top of the chromatogram, remove it from the eluting chamber and mark the position of the solvent front with a pencil. This must be done quickly, because the solvent evaporates rapidly. Allow the chromatogram to dry in air; record the positions and color of the bands in your notebook. The chromatogram will generally appear like that shown in Figure 11.5.

H. Development of Chromatogram

After the chromatogram is dry, attach it to the paper clip hook, and hang it in the developing chamber containing NH$_3$. The apparatus at this point should look like that shown in Figure 11.6.

NOTE: The chromatogram should not touch the solution in the bottom of the test tube.

The object of the development step is to bathe the chromatogram in NH$_3$ vapors which may cause the formation of complex compounds with characteristic colors. The development should take about 30 seconds. Record in your notebook any color changes which may occur in the bands on your chromatogram.

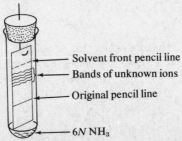

Fig. 11.6 *The eluted chromatogram is developed by exposing it in an atmosphere of NH$_3$ solution.*

Table 11.1

Species	Color
$Fe(H_2O)_6^{3+}$	yellow
$FeCl_3$	yellow
$Co(H_2O)_6^{2+}$	red
$CoCl_2$	blue
$CoCl_2(NH_3)_2$	rose-red or bluish-violet; gray-green when wet
$Cu(H_2O)_6^{2+}$	pale blue
$CuCl_2(H_2O)_2$	green
$CuCl_2$	yellow; pink in acetone
$Cu(NH_3)_4(H_2O)_2^{2+}$	dark blue
$Ni(H_2O)_6^{2+}$	green
$NiCl_2(H_2O)_2$	pale green
$[Ni(NH_3)_4(H_2O)_2]^{2+}$	purple-violet
$Mn(H_2O)_6^{2+}$	slightly colored
$MnCl_2$	slightly colored
$Mn(OH)_2$	red-brown
MnO	red-brown

III. DATA ANALYSIS

Transition metal ions form coordination compounds which exhibit characteristic colors depending upon the ligands present. The predominating ligands in the eluting solution are Cl^- (from HCl) and H_2O. The development provides a source of NH_3 ligands. Thus, it is possible that the transition-metal ions in your unknown may contain either Cl^- or H_2O ligands or a mixture of these species. When exposed to ammonia, the ions may contain mixtures of Cl^-, H_2O, and NH_3 as ligands. Table 11.1 contains a list of the possible complex compounds and the colors of the transition metal ions that may be a part of your unknown. This information, plus your observations on the freshly eluted chromatogram and the developed chromatogram should be sufficient to identify the cations in your unknown. For example, a green band in your chromatogram which turned violet upon exposure to NH_3 indicates the presence of nickel. If your interpretation is ambiguous, you can prepare a chromatogram of a solution containing a known ion (or an ion you think might be present) and compare the results.

I. R_f Values

The rate at which an ion moves during elution is characteristic of that ion. The rates at which ions move are generally related to the movement of the solvent front. A measure of the rate at which an ion moves is called its R_f *value*. Each ion has its own characteristic R_f value in a given solvent using a given paper. The R_f value of an ion is defined by the expression

$$R_f = \frac{\text{distance ion moves}}{\text{distance solvent front moves}} \tag{2}$$

Thus, if an ion moves as rapidly as the solvent front, its R_f value is 1.0; generally, the ions move less rapidly than the solvent and R_f values are less than one.

K. Sample Calculation of R_f Values

Assume your chromatogram looks like the one shown in Figure 11.7. Measure (in cm) the farthest distance each ion has traveled from the starting line and the total distance from the starting line to the solvent front. Assume the data are as given in Figure 11.7. Using Equation 2 the R_f values of the ions are calculated as

$$R_f = \frac{3}{10} = 0.3$$

$$R_f = \frac{5}{10} = 0.5$$

$$R_f = \frac{9}{10} = 0.9.$$

The report for this experiment consists of naming the ions present in your unknown and calculating their R_f values.

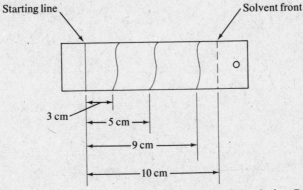

Fig. 11.7 *Measurements to be made on an eluted chromatogram to calculate R_f values for various bands.*

NAME _____ DATE _____

SECTION _____

SELF-STUDY QUESTIONS

1. Distinguish between *ab*sorption and *ad*sorption.

2. In chromatography, what is the difference between elution and development?

3. What two properties of ions permit you to identify them in a chromatographic experiment?

4. In Table 11.1 it is indicated that a possible species in the developed chromatogram is $Mn(OH)_2$. Give balanced equations which explain the presence of this species on the developed chromatogram.

5. Two colored ions exhibit R_f values of 1.00 and 0.95, respectively, in a given solvent.

a. Give a sketch of what you would expect the chromatogram to look like.

b. Why would chromatography be a poor basis for the identification of these ions?

c. What variables could you attempt to change to improve the basis for identification of these ions?

Introduction to the Qualitative Analysis of Cations

Experiments 12, 13, 14, and 15 are devoted to a study of qualitative analysis—the methods used to identify the species present in a mixture. In these experiments you will be asked to analyze solutions containing metallic cations. The cations with which we are concerned include Ag^+, Pb^{2+}, Bi^{3+}, Cu^{2+}, Cd^{2+}, As^{5+}, Fe^{3+}, Co^{2+}, Ni^{2+}, Cr^{3+}, Ba^{2+}, and Ca^{2+}. Some analytical schemes include considerably more cations than are in this list, but the present—abbreviated—list is sufficient for our purposes.

Ideally this task could be accomplished if there existed specific reagents for each cation present which would produce a characteristic precipitate or color with each cation. Unfortunately such an ideal collection of reagents does not exist and we must use a method which separates cations into groups containing a smaller number of cations than the original group. These smaller groups can then be separated to yield observations on individual cations in an environment where they are free of other cations. Thus, the whole plan of qualitative analysis involves a series of separations and identifications.

The members of each group and the conditions under which that group is separated are given in the table below.

After separation, the cations within a group are further resolved by means of a

Table 1

Group Number	Members	Separation Conditions
I	Ag^+, Pb^{2+}	Precipitated as chlorides in strong acid
II	Cu^{2+}, Cd^{2+} Bi^{3+}, As^{5+}	Precipitated as sulfides in a weak acid solution
III	Cr^{3+}, Ni^{2+}, Co^{2+}	Precipitated as sulfides or hydroxides in basic solution
IV	Ba^{2+}, Cu^{2+}	Remain in solution after ions in Groups I–III are precipitated

series of chemical reactions into sets of soluble and insoluble fractions; this process is continued until the resolution is sufficient to allow identification of each cation by one or more tests specific to that ion. In addition to precipitations, we will employ acid-base reactions, complex ion formation, oxidation-reduction, and combinations of these kinds of reactions to accomplish the analyses.

Flow charts are convenient methods to represent general chemical procedures. In the figure below is the flow chart for the scheme used to separate the cations into groups. Imagine that the unknown solution contains all the possible cations in the scheme as you read the flow chart. The solution is treated sequentially with reagents which precipitate a group of cations, the precipitate is separated from the solution, and the remaining solution is treated with another reagent.

Your instructor will assign an unknown in each of the separate groups—an unknown which consists of ions in all groups (a general unknown) or both types.

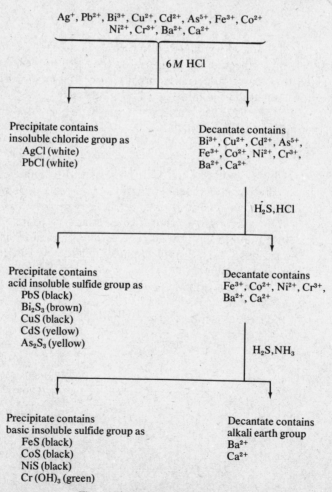

Ag$^+$, Pb^{2+}, Bi^{3+}, Cu^{2+}, Cd^{2+}, As^{5+}, Fe^{3+}, Co^{2+}
Ni^{2+}, Cr^{3+}, Ba^{2+}, Ca^{2+}

6 M HCl

Precipitate contains
insoluble chloride group as
    AgCl (white)
    PbCl (white)

Decantate contains
Bi^{3+}, Cu^{2+}, Cd^{2+}, As^{5+},
Fe^{3+}, Co^{2+}, Ni^{2+}, Cr^{3+},
Ba^{2+}, Ca^{2+}

H$_2$S,HCl

Precipitate contains
acid insoluble sulfide group as
    PbS (black)
    Bi$_2$S$_3$ (brown)
    CuS (black)
    CdS (yellow)
    As$_2$S$_3$ (yellow)

Decantate contains
Fe^{3+}, Co^{2+}, Ni^{2+}, Cr^{3+},
Ba^{2+}, Ca^{2+}

H$_2$S,NH$_3$

Precipitate contains
basic insoluble sulfide group as
    FeS (black)
    CoS (black)
    NiS (black)
    Cr (OH)$_3$ (green)

Decantate contains
alkali earth group
Ba^{2+}
Ca^{2+}

Fig. 1 *The group separation scheme.*

LABORATORY PROCEDURES

Good laboratory technique in qualitative analysis is important if results like those described in the procedures are to be obtained. Careless work will produce precipitates of the wrong color or precipitates where you should get solutions, and in general will make you less successful than you might otherwise have been.

The analyses described here will be carried out on the semimicro level. The amounts of chemicals used are neither large (grams) nor very small (micrograms). The sample solutions are typically about 0.1 M, and the volume of those solutions will be about 1 ml, which means that there will be about 10 mg of solute in the sample. The identification tests will be ineffective if you do not have at least 1 mg of solute present. The experiments must be carried out carefully to avoid losing the major portion of the components during the analysis.

MEASURING OUT LIQUIDS

Typically a procedure calls for adding 1 ml of a reagent to a solution. This does not mean that you should use a small graduated cylinder to measure the reagent. There are several ways you can learn to *estimate* 1 ml of liquid. You can learn to visualize an increase in liquid level of 1 ml of liquid in a semimicro test tube by pouring 1 ml of water at a time from a small graduated cylinder. Or you can determine the average number of drops per milliliter by using a medicine dropper. (Typically droppers deliver 10 to 15 drops per milliliter.)

Never contaminate a reagent solution by dipping your own medicine dropper or pipette into it. If you must have a larger volume of reagent than is conveniently handled by the dropper (and this should be a very rare occurrence), pour the reagent into a clean beaker and then dispense it with your own *clean* dropper.

MEASURING OUT SOLIDS

Dispensing an approximate amount of solid can be accomplished in much the same ways as those for liquids. A small amount of solid is usually added from the tip of a spatula, the amount being estimated by the volume it occupies. (The assumption is that most solid substances have nearly the same densities so that equal volumes will give approximately equal weights.) To get experience in judging the amount of solid from its volume, use the following procedure. Weigh a small empty beaker on a sensitive balance, and add small portions of a solid substance from the tip of a spatula. You should be able to reach the point where you can deliver 0.1 ± 0.02 g of solid from a spatula quite accurately.

STIRRING RODS

A glass stirring rod is a very useful tool in analyses; you need it for nearly every step in each procedure. After each time it is used, of course, it has the solution being treated on it, and therefore must be cleaned before it is used again. Keep a 250 or 400 ml beaker full of distilled water and, after you have used a stirring rod, swirl it around in the water to clean it; then leave it in the water. Solutes and solids eventually accumulate in the beaker, but they amount to only tiny traces when diluted by the water. There will be no significant contamination if you change the water occasionally.

PRECIPITATION

One of the most common reactions in qualitative analysis is precipitation from solution. The indicated amount of precipitating reagent is added to the sample solution and stirred well with a glass stirring rod. Heat the solution in the water bath if so directed. Since some precipitates form slowly, they must be given enough time to form completely. Frequently, students do not mix reagents thoroughly enough. When you think the precipitation is complete, centrifuge the mixture and, before decanting the liquid, add a drop or two of the precipitating reagent, just to be sure that you've added enough precipitant the first time. The directions usually specify enough reagent to furnish an excess, but it is good insurance to check.

SEPARATION OF SOLID FROM SOLUTIONS

One of the reasons for working at the semimicro level is that small centrifuges can be used conveniently to separate a precipitate from a solution. The common laboratory centrifuge will accept a small semimicro test tube holding about 8 ml; such tubes (13 mm × 100 mm) are large enough to hold the reaction mixtures found in qualitative analysis.

   In general, to centrifuge a sample, put the test tube containing the precipitate and solution into one of the locations in the centrifuge and another test tube containing a similar amount of water on the opposite side of the rotor. Turn on the centrifuge and let it run for about 30 seconds, during which time most precipitates will settle to a compact mass in the bottom of the tube. If your sample is still suspended, centrifuge it again; if this still does not work, it may be helpful to heat the sample in the water bath for a few minutes, thereby promoting formation of larger crystals of solid, which tend to centrifuge out more easily.

WASHING SOLIDS

The liquid decanted from the test tube after centrifuging out a solid does not in general contain any of the solid and may be used directly in a following analysis step. The solid remaining in the tube, however, has residual liquid around it. Since this liquid contains ions that may interfere with further tests on the solid, they must be removed by diluting the liquid with a wash liquid (often water) which does not interfere with the analysis. To wash the solid, add the indicated amount of wash liquid to the test tube and mix well with a glass stirring rod to disperse the solid well in the wash liquid. After mixing thoroughly, centrifuge out the solid and decant the wash liquid, which usually may be discarded. In key separations, it is best to perform the washing operation twice because an uncontaminated precipitate tends to give much better results than one mixed with even a trace of contamination.

HEATING

Many reactions are best carried out when the reagents are hot. Never heat a test tube containing a reaction mixture directly on a Bunsen flame. It is more convenient and much more considerate of your colleagues to heat the test tube in a water bath. A 150 ml beaker containing about 100 ml of water makes an adequate water bath.

When you are following a procedure in which you need to heat a solution, keep the water bath hot or boiling by using a small flame to heat the beaker on a piece of asbestos-covered wire gauze. This way, the bath is ready whenever required. The test tubes can be put directly into the bath, supported against the wall of the beaker. A mixture under study can be brought to approximately 100°C without boiling it. If the sample is heated in an open flame, it inevitably bumps out of the tube and onto the laboratory bench, or onto you or someone else. To repeat: do not heat test tubes containing liquids over an open flame.

EVAPORATING A LIQUID

In some cases, it will be necessary to decrease the volume of a liquid to concentrate a species or to remove a volatile reagent. In some cases, this will be necessary because the reaction of interest will proceed readily only in a boiling solution. When you must decrease the volume of a liquid, perform the evaporation in a small 30 ml beaker on a square of asbestos-covered wire gauze and use a small Bunsen flame, judiciously applied to maintain controlled, gentle boiling. Since the volumes involved usually are of the order of 3 or 4 ml, they can easily be overheated.

If you are supposed to stop the evaporation at a volume of 1 ml or so, don't heat to dryness or you may decompose the sample or render it inert. Since concentrated solutions or slurries tend to bump, it is often helpful to encourage smooth boiling by scratching the bottom of the beaker with a stirring rod as the boiling proceeds. Good judgment in boiling down a sample is important.

Occasionally the liquid that is evaporated is highly acidic (HCl or HNO$_3$) and the boiling causes evolution of noticeable amounts of toxic gases into the laboratory. In some cases the amounts are small enough to be ignored. If you notice that bothersome vapors are escaping from a boiling mixture, transfer the operation to the hood.

TRANSFERRING A SOLID

Sometimes it is necessary to transfer a solid from the beaker in which it was prepared to a test tube for centrifuging, or from the test tube to the beaker. The amount of solid involved is never large and is about 50 mg at most. Such transfers can be made in the presence of a liquid, which serves as a carrier. When you are ready to make the transfer, stir the solid well into the liquid to form a slurry and then, without delay, pour the slurry into the other container. The transfer is not quantitative but, if done properly, you can move 90 percent of the sample to the other container. Forget about the rest. In general, do not attempt to transfer wet solids with a spatula because it is not easy, and all too often the spatula reacts with the liquid present, contaminating it.

ADJUSTING THE *p*H OF A SOLUTION

One of the most important experimental variables controlling chemical reactions is the *p*H of the solution. Frequently it is necessary to make a basic solution acidic, or vice versa, in order to make a desired reaction take place. For example, if you are directed to add 6 *M* HCl to a mixture until it is acidic, you should proceed as follows: Knowing how the solution was prepared, make a quick mental calculation of about how much acid is needed—1 drop, 1 ml, perhaps more. Then add the acid drop by drop, until you think the *p*H is about right. Mix well with a stirring rod and then touch the end of the rod to a piece of blue litmus paper on a piece of paper towel or filter paper. If the color does not change, add another drop or two of acid, mix, and test again. Frequently the system changes its character at the neutral point; a precipitate may dissolve or form, or the color may change. In any event, add enough acid so that, after mixing, the litmus paper turns red when touched with the stirring rod. Similarly, if you are told to make a solution basic with 6 *M* NH$_3$ or 6 *M* NaOH, add the reagent, drop by drop, until the solution, after being well mixed, turns red litmus blue. Adjustment of *p*H is not difficult, but it must be done properly if the desired reaction is to occur.

EXPERIMENT

12

Qualitative Analysis of the Group I Cations: Ag^+, Pb^{2+}

PROPERTIES OF THE GROUP I CATIONS

Ag^+

Only a few water soluble salts of silver are known: the nitrate is certainly the most common. Most of the insoluble silver salts dissolve in cold $6\,M$ HNO_3, the main exceptions being the silver halides, AgSCN, and Ag_2S. Silver ion forms many stable complexes; of these, the best known is probably the $Ag(NH_3)_2^+$ ion. This complex is sufficiently stable to be produced when AgCl or AgSCN is treated with $6\,M$ NH_3; the reaction which occurs is useful for dissolving those solids. AgBr and AgI are less soluble than AgCl; AgBr will go into solution in $15\,M$ NH_3, but AgI is so insoluble that it will not dissolve in NH_3. The silver thiosulfate complex ion, $Ag(S_2O_3)_2^{3-}$ is extremely stable and is important in photography, where it is formed in the "fixing" reaction in which AgBr is removed from the developed negative.

Pb^{2+}

Lead nitrate and acetate are the only well-known soluble lead salts. Lead chloride is not nearly as insoluble in water as silver chloride and becomes moderately soluble if the water is heated. $PbSO_4$ is one of the few insoluble sulfates. Lead forms a stable hydroxide complex ion and a weak chloride complex. Although lead ordinarily has an oxidation number of $+2$, there are some Pb(IV) compounds, of which the most common is PbO_2 (brown); this compound is insoluble in most reagents, but it will dissolve in $6\,M$ HNO_3 if some H_2O_2 has been added.

Some of the general solubility characteristics of the Group I cations are listed in Table 12.1. You should try to familiarize yourself with the information in this table. Thus, for example, the entry under OH^- and Pb^{2+} (which corresponds to $Pb(OH)_2$) indicates that lead hydroxide is a colorless (white) substance, soluble in a solution containing a good complex agent (C), and soluble in acid (A). The last line in the table indicates the species that form good complex ions with the cation; for example, NH_3 and $S_2O_3^{2-}$ form complex ions with Ag^+.

Table 12.1
Solubility Properties of the Group I Cations

	Ag^+	Pb^{2+}
Cl^-	C, A^+ (white)	HW, C, A^+ (white)
OH^-	C, A (brown)	C, A (white)
SO_4^{2-}	S^-, C (white)	C (white)
CrO_4^{2-}	C, A (dk red)	C (yellow)
CO_3^{2-}, PO_4^{3-}	C, A (white)	C, A (white)
S^{2-}	O (black)	O (black)
Complexes	NH_3, $S_2O_3^{2-}$	OH^-

Note: S means soluble in water, >0.1 mole/l. S^- means slightly soluble in water, ~0.01 mole/l. HW means soluble in hot water. A means soluble in acid (6 M HCl or other nonprecipitating, nonoxidizing acid). A^+ means soluble in 12 M HCl. O means soluble in hot 6 M HNO_3. C means soluble in solution containing a good complexing ligand.

SEPARATION OF THE GROUP I CATIONS

The first ions which are determined are Ag^+ and Pb^{2+}. Of all the common cations, only these form insoluble chlorides. These cations comprise Group I in the scheme and are separated from a general salt solution by precipitation of their chlorides under acidic conditions (Fig. 1, p. 122). The chloride precipitate is then analyzed for the possible presence of silver and lead on the basis of the characteristic properties of those cations.

GENERAL SCHEME OF ANALYSIS

The procedure for the qualitative analysis of Group I is outlined in the flow chart given in Figure 12.1, and the detailed directions are given in Table 12.2.

DATA ANALYSIS

Construct a flow chart showing the behavior of your unknown when subjected to analysis.

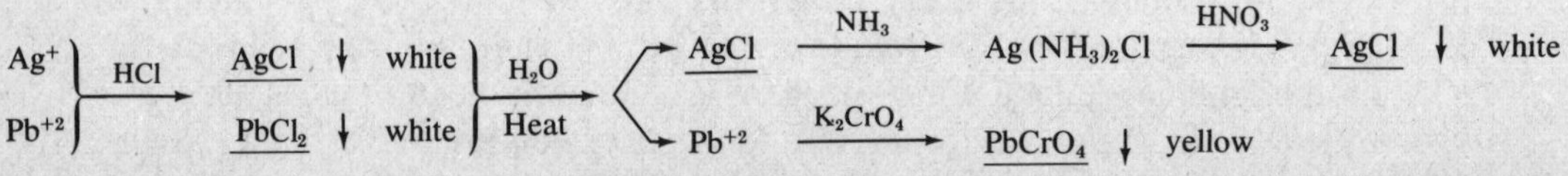

Fig. 12.1 *Group I flow chart.*

Table 12.2
Group I Flow Chart

To 15 drops of the sample add 2 drops of 6 M HCl, mix thoroughly by stirring with a glass stirring rod. Cool in ice bath for 3 minutes and centrifuge. Test for complete precipitation by adding 1 drop of 6 M HCl to the mixture and observing the formation of additional precipitate. If precipitation is complete, decant. Save the decantate for subsequent analysis of other groups.

Precipitate: AgCl, $PbCl_2$
Wash the precipitate with 5 drops of cold water and discard wash liquid. To the precipitate add 10 drops of water and heat in a boiling water bath for 3 minutes; centrifuge and decant. Repeat the extraction with 5 more drops of water. Centrifuge while still hot, and combine both decantates.

Solution: Save for analysis of other groups (see Fig. 1, p. 122 and Table 1, p. 121).

Residue: AgCl
Dissolve in 10 drops of 6 M NH₃ solution. Acidify the solution with 6 M HNO₃. A white precipitate of AgCl confirms the presence of Ag⁺¹.

Solution: Pb^{+2}
Add 2 drops of 1 M K_2CrO_4. A yellow precipitate of $PbCrO_4$ confirms the presence of Pb^{+2}.

In a solution saturated with H_2S,

$$K_{(H_2S)} = [H^+]^2 [S^{2-}] = 1.3 \times 10^{-21}$$

Prior to the addition of thioacetamide, the hydrogen ion concentration must be carefully controlled. Evaporation of the sample to a moist residue ensures removal of most of the acid present in the unknown so that a known concentration of acid can be introduced. The addition of 10 drops of 0.6 M HCl and 10 drops of thioacetamide to the moist solid gives a solution with a hydrogen ion concentration of 0.3 M. The hydrolysis of thioacetamide in acidic medium yields H_2S according to the following equation:

$$CH_3CSNH_2 + 2H_2O + H^+ \longrightarrow H_2S + CH_3COOH + NH_4^+$$

Since $[H^+] = 0.3\,M$, the

$$[S^{2-}] = \frac{K_{(H_2S)}}{[H^+]^2} = \frac{1.3 \times 10^{-21}}{(3 \times 10^{-1})^2} = 1.4 \times 10^{-20}\,M$$

This sulfide concentration is high enough to give an effective precipitation of the acid insoluble group sulfides while the more soluble sulfides of the following group (Group III) are not precipitated under these conditions.

Since the solubility products of the Group II sulfides are very low, they precipitate in the presence of even extremely low concentrations of S^{2-} ion. The precipitation is carried out at a pH of 0.5, where $[S^{2-}]$ is only about $1 \times 10^{-21}\,M$. The values of the solubility products of some common metallic sulfides are given in Table 13.2. Under the conditions of the precipitation, the metallic cations are all about 0.01 M; therefore, $[M^{2+}][S^{2-}] = (1 \times 10^{-2})(1 \times 10^{-21}) \approx 1 \times 10^{-23}$, which means that those sulfides in the left column of Table 2, $K_{sp} < 10^{-23}$, will precipitate, whereas those in the right column, $K_{sp} \geq 10^{-23}$, will not. This difference in sulfide solubilities is the basis of the separation of the Group II cations (left column) from Group III cations (right column).

Table 13.2
Solubility Products of the Group II and Group III Sulfides

Group II		*Group III*	
CuS	1×10^{-36}	CoS	1×10^{-21}
CdS	1×10^{-26}	NiS	1×10^{-22}
PbS	1×10^{-27}	ZnS	1×10^{-23}
		FeS	6×10^{-18}

GENERAL SCHEME OF ANALYSIS

The procedure used for the qualitative analysis of Group II is outlined in the flow chart given in Figure 13.1, and the detailed directions are given in Table 13.3.

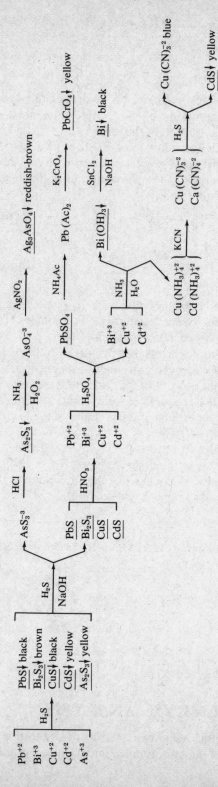

Fig. 13.1 *Group II flow chart.*

Table 13.3

Group II Flow Chart

Transfer the solution obtained from the insoluble chloride group separation to an evaporating dish and evaporate to a moist residue. Cool and take up in 10 drops of $0.6M$ HCl. Transfer the solution to a small test tube and add 10 drops of $1M$ thioacetamide. Heat in a hot water bath for 5 minutes. Add 20 drops of water and heat for an additional 5 minutes. Centrifuge and test the solution for complete precipitation by the addition of 1 drop of thioacetamide. If more precipitate forms, repeat the heating process. If precipitation is complete, decant and wash the precipitate with 10 drops of $0.1M$ HCl and discard wash liquid. Save the decantate for the analysis of the following groups.

Precipitate: PbS, Bi_2S_3, CuS, CdS, As_2S_3.
To the residue add 10 drops of $6M$ NaOH and 8 drops of $1M$ thioacetamide.
Heat in a water bath for 5 minutes. Centrifuge and decant.

Solution: Save for the analysis of the following groups (see Experiment 14).

Precipitate: PbS, Bi_2S_3, CuS, CdS
Dissolve with 10 drops of $6M$ HNO_3 and heat to expel nitrogen oxide fumes. Cool, centrifuge, and discard any sulfur which may be floating on the solution.

Solution: AsS_3^{-3}. Acidify with $6M$ HCl. Centrifuge and discard the decantate. Residue: As_2S_3. Add 5 drops of 3 percent H_2O_2 and 6 drops of $6M$ NH_3 solution. Transfer to an evaporating dish; heat to a moist residue. Cool, then add 4 drops of $0.2M$ $AgNO_3$. A reddish-brown precipitate confirms the presence of arsenic.

Solution: Pb^{+2}, Bi^{+3}, Cu^{+2}, Cd^{+2}
Place the solution in an evaporating dish and add 3 drops of $6M$ H_2SO_4, evaporate under the hood until the appearance of white SO_3 fumes. Cool, and add 20 drops of water. Transfer the solution and solid, if any, to a small test tube. Centrifuge and decant.

Precipitate: $PbSO_4$
Dissolve in 10 drops of $1M$ NH_4Ac and 2 drops of $6M$ HAc. The dissolution may require some heating. Add 2 drops of $1M$ K_2CrO_4. A yellow precipitate of $PbCrO_4$ confirms the presence of Pb^{+2}.

Solution: Bi^{+3}, Cu^{+2}, Cd^{+2}
Make the solution distinctly basic by the addition of $15M$ NH_3 solution. Centrifuge and decant.

Precipitate: $Bi(OH)_3$
Add 6 drops of $6M$ NaOH and 3 drops of $0.5M$ $SnCl_2$. A black precipitate of Bi confirms the presence of Bi^{+3}.

Solution: $Cu(NH_3)_4^{+2}$, $Cd(NH_3)_4^{+2}$
A blue solution at this step confirms the presence of Cu^{+2}. Divide the solution into two parts:

First part: Acidify with $6M$ HAc. Add 3 drops of $0.2M$ $K_4Fe(CN)_6$. A red precipitate of $Cu_2Fe(CN)_6$ confirms the presence of Cu^{+2}.

Second part: Add $0.2M$ KCN until the blue color disappears. Then add 6 drops of $1M$ thioacetamide. Heat in a hot water bath for 3 minutes. A yellow precipitate of CdS confirms the presence of Cd^{+2}.

DATA ANALYSIS

Prepare a flow chart showing the behavior of your unknown when subjected to analysis.

NAME   **DATE**

SECTION

SELF-STUDY QUESTIONS

1. Write balanced equations for the following processes:
a. the dissolution of As_2S_3;
b. the dissolution of Bi_2S_3;
c. the confirmatory test for arsenic;
d. the confirmatory test for cadmium.

2. Devise a test for identifying the following ions:
a. Cu^{2+} in the presence of Bi^{3+};
b. As^{3+} in the presence of Pb^{2+};
c. Cd^{2+} in the presence of Cu^{2+};
d. Bi^{3+} in the presence of Cd^{2+}.

3. Assume you have a solution which contains only one cation in Group II. Give a flow chart describing the minimum number of steps necessary to identify the ion.

EXPERIMENT

14

Qualitative Analysis of the Group III Cations: Fe^{2+}, Co^{2+}, Ni^{2+}, Cr^{3+}

PROPERTIES OF THE GROUP III CATIONS

Fe^{3+}

Iron in its compounds is ordinarily found in the +2 (ferrous) or the +3 (ferric) states. The latter is more common, since most ferrous compounds oxidize in the air, particularly if water is present. Iron(II) compounds are usually found as hydrates and are light green. Iron(III) salts are also ordinarily obtained as hydrates and are often yellow or orange. Both Fe^{2+} and Fe^{3+} form many complexes; perhaps the most stable are those with cyanide, $Fe(CN)_6^{4-}$ and $Fe(CN)_6^{3-}$. The $FeSCN^{2+}$ ion has a very characteristic deep red color. Metallic iron is a good reducing agent, dissolving readily in $6M$ HCl with evolution of hydrogen. Conversion of Fe(II) to Fe(III) or the reverse is easily accomplished by common oxidizing agents (air, H_2O_2 in acid) or by reducing agents such as H_2S, Sn^{2+}, I^-.

Co^{2+}

Cobalt(II) salts in water solution are characteristically pink, the color of the hydrated cobalt ion, $Co(H_2O)_6^{2+}$. The pink color is usually too delicate to be used to characterize the ion. Cobalt(II) forms several complex ions. When heated to boiling, the color of cobalt chloride solutions turns from pink to blue. The color change occurs because of a change in coordination number; the pink form arises from octahedrally coordinated Co(II) whereas the blue form is tetrahedrally coordinated Co(II). Cobalt sulfide does not dissolve readily in $6M$ HCl even when heated. Co(II) reacts with thiocyanate solutions to form a blue complex, $Co(SCN)_4^{2-}$, whose stability is much greater in ethanol than in water. Addition of KNO_2 to solutions of Co^{2+} produces a characteristic yellow precipitate of $K_3Co(NO_2)_6$ in which the metal has been oxidized (Eq. 1).

$$Co^{2+}(aq) + 7NO_2^-(aq) + 3K^+(aq) + 2H^+(aq) \longrightarrow$$
$$K_3Co(NO_2)_6(s) + NO(g) + H_2O \qquad (1)$$

Under strongly oxidizing conditions Co(II) can be converted to Co(III), which has a stability that is enhanced in a complex species like $Co(NH_3)_6^{3+}$, or an insoluble substance such as the yellow cobaltinitrite produced in the reaction above.

Ni^{2+}

Nickel salts, like those of most of the other members of Group III, are typically colored, the hydrates being green, which is the color of the $Ni(H_2O)_6^{2+}$ complex ion. Nickel forms several complex ions, many of which have characteristic colors; the blue $Ni(NH_3)_6^{2+}$ ion is perhaps the most common of these. Nickel sulfide, NiS, when first precipitated tends to be colloidal and difficult to settle by centrifuging; it is not very soluble in $6M$ HCl, even though it cannot be precipitated from acidic solutions. Nickel forms a very characteristic rose-red precipitate with an organic reagent called dimethylglyoxime; this product is useful in the quantitative analysis of nickel.

Cr^{3+}

Chromium is ordinarily encountered in either the +3 or the +6 oxidation state. The common cation is Cr^{3+}, which forms several complexes, all of which are colored; thus, $Cr(H_2O)_6^{3+}$ is reddish-violet in solution. The chromium-containing precipitate obtained in the Group III precipitation scheme is the hydroxide, $Cr(OH)_3$, rather than the sulfide. Chromium(III) can be oxidized to Cr(VI) with several oxidizing agents such as ClO_3^- in $16M$ HNO_3, H_2O_2 in $6M$ NaOH, and ClO^- in $6M$ NaOH. Chromium is usually identified in the form of a Cr(VI) species in neutral or basic solutions; Cr(VI) exists as the bright yellow chromate, CrO_4^{2-}, ion. If acid is added to this species, orange dichromate ion, $Cr_2O_7^{2-}$, is produced (Eq. 2).

$$2CrO_4^{2-}(aq) + 2H^+(aq) = Cr_2O_7^{2-}(aq) + H_2O \tag{2}$$

If H_2O_2 is added to a solution containing $Cr_2O_7^{2-}$ ion, a transitory blue color, due to CrO_5 is observed. This reaction forms the basis for the confirmatory test for chromium.

Some of the general solubility properties of Group III cations are given in Table 14.1.

SEPARATION OF THE GROUP III CATIONS

The sulfides of this group are much more soluble than the sulfides of Group II. Therefore a higher concentration of sulfide ion is required to precipitate the ions of

Table 14.1
Solubility Properties of the Group III Cations

	Fe^{3+}	Co^{2+}	Ni^{2+}	Cr^{3+}
Cl^-	S	S	S	S
OH^-	A (red-brown)	A (tan)	A (green)	C, A (green)
SO_4^{2-}	S	S	S	S
CrO_4^{2-}	A (tan)	C, A (brown)	S	A (tan)
CO_3^{2-}, PO_4^{3-}	A (tan)	C, A (violet)	C, A (green)	A (green)
S^{2-}	D (black)	A^+, O^+ (black)	A^+, O^+ (black)	D (green)
Complexes	—	NH_3^*	NH_3	OH^*

*Although cobalt forms ammonia complexes, some of them are insoluble, so that tests for solubility should be made for mixtures of interest. Chromium hydroxide coprecipitated with other hydroxides may be insoluble in 6M NaOH.

Note: S means soluble in water, > 0.1 mole/l. A means soluble in acid (6M HCl or other non-precipitating, nonoxidizing acid). O^+ means soluble in hot aqua regia. C means soluble in solution containing a good complexing ligand. D means unstable, decomposes.

Group III. From the previous discussion it was shown that the sulfide ion concentration is governed by the pH of the solution according to Equation 3.

$$[S^=] = K_{(H_2S)}/[H^+]^2 \tag{3}$$

In order to increase the sulfide concentration, we must increase the pH of the solution; in the procedure we add NH_3 solution. Thioacetamide is the group precipitation reagent. The hydrolysis of thioacetamide in a basic medium yields sulfide ion according to Equation 4.

$$CH_3CSNH_2 + 3OH^- \longrightarrow CH_3COO^- + S^= + NH_3 + H_2O \tag{4}$$

The only cation which is not precipitated as a sulfide in this group is Cr^{+3}. Chromium hydroxide is so insoluble that it will precipitate even in the presence of the low concentration of hydroxide ion which arises in an aqueous ammonia solution (Eq. 5).

$$NH_3 + H_2O \rightleftharpoons NH_4^+ + OH^- \tag{5}$$

GENERAL SCHEME OF ANALYSIS

The procedure used for the qualitative analysis of Group III is outlined in the flow chart given in Figure 14.1 and the detailed directions are given in Table 14.2.

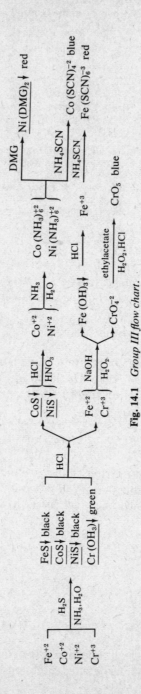

Fig. 14.1 *Group III flow chart.*

Table 14.2
Group III Flow Chart

To the solution obtained from previous group separation add 5 drops of $1M$ thioacetamide and heat in a water bath for 5 minutes. Make the solution slightly basic by adding $15M$ NH_3 solution; then add 2 drops in excess. Heat for 5 minutes. Centrifuge and test the solution for complete precipitation by adding more NH_3 solution. If more solid forms, add 2 drops of thioacetamide and repeat the heating process. If precipitation is complete, decant.

Solution: Save for the analysis of the Alkaline Earth Group (see Fig. 1, p. 121 and Table 1, p. 122).

Precipitate: NiS, CoS, FeS, $Cr(OH)_3$
Wash with 10 drops H_2O and discard wash liquid. Treat the precipitate with 10 drops of $1M$ HCl. Stir, centrifuge, and separate the mixture immediately. Repeat the treatment with 5 more drops of HCl, and save the combined decantates.

Residue: CoS, NiS
Dissolve in 9 drops of $12M$ HCl and 3 drops of $16M$ HNO_3. Separate any sulfur that may form, then heat the solution in a bath to expel oxides of nitrogen.

Solution: Co^{+2}, Ni^{+2}
Cool, and add $6M$ NH_3 solution until the solution is basic. Divide into two parts.

Part 1. Add 3 drops of dimethylglyoxine. The formation of a pink or red precipitate confirms the presence of Ni^{+2}.

Part 2. Make the solution acidic to litmus with $6M$ HCl. Add 5 drops of $1M$ NH_4SCN and 2 ml of acetone. A blue solution confirms Co^{+2}. If a red color is observed Fe^{+3} is interfering. Add $1M$ NaF until the red color disappears.

Solution: Cr^{+3}, Fe^{+2}
Add 10 drops of $6M$ NaOH and 6 drops of 3 percent H_2O_2. Heat in a water bath for 5 minutes. Cool, centrifuge, and decant.

Precipitate: $Fe(OH)_3$
Dissolve in 5 drops of $6M$ HCl, then add 3 drops of $1M$ NH_4SCN. A red solution confirms the presence of Fe^{+3}.

Solution: CrO_4^{-2}
Add 8 drops of ethylacetate. Prepare a mixture of 5 drops of $6M$ HCl and 5 drops of 3 percent H_2O_2, and add it dropwise to the previous solution. The appearance of a blue color in the organic layer confirms the presence of Cr^{+3}.

DATA ANALYSIS

Construct a flow chart showing the behavior of your unknown when subjected to analysis.

SELF-STUDY QUESTIONS

1. Write balanced ionic equations for the following processes:
a. the confirmatory test for Co^{2+};
b. the group precipitation of Cr^{3+};
c. the confirmatory test for chromium;
d. the separation of Fe^{2+} from Cr^{3+}.

2. What is the single experimental parameter which can be varied to separate Cu^{2+} and Fe^{2+} using S^{2-} as a precipitating reagent? Give balanced equations to support your description.

3. What is the S^{2-} ion concentration in an aqueous solution of $pH = 8$ saturated with H_2S?

4. Explain why $Cr(OH)_3$ precipitates in the basic insoluble sulfide group.

5. Each of three test tubes contains one of the following ions. Describe how you would decide which ion is in which test tube.
   a. Co^{2+}, Cd^{2+}, Ag^+;
   b. Pb^{2+}, Cr^{3+}, Cu^{2+};
   c. Ni^{2+}, Fe^{3+}, Pb^{2+}.

EXPERIMENT

15

Qualitative Analysis
of the Group IV
Cations: Ba^{2+}, Ca^{2+}

PROPERTIES OF THE GROUP IV CATIONS

Ba^{2+}

Barium exists in its compounds essentially only in the +2 state and is not reducible to the metal in aqueous systems. Most of the common barium salts are soluble in water or dilute strong acids and are colorless in solution. The main exception is the sulfate $BaSO_4$, a finely divided white powder that is essentially insoluble in all common reagents. Barium chromate, $BaCrO_4$, is insoluble in acetic acid but will dissolve in solutions of strong acids. The oxalate and phosphate are soluble in acidic systems at a pH of 3 or less. Barium hydroxide is a strong base and is moderately soluble in water ($\sim$0.2 mole/l).

Ca^{2+}

Calcium in its compounds occurs in the +2 state, and the ion is very difficult to reduce to the metal. Calcium salts are typically soluble in water or dilute acids. Like barium, calcium does not form many common complex ions. Although the hydroxide is less soluble than $Ba(OH)_2$, it will not precipitate in ammonia. Calcium oxalate, CaC_2O_4, is white and not appreciably soluble in acetic acid, but it will dissolve in dilute strong acid solutions.

Some of the general solubility properties of the Group IV cations are given in Table 15.1.

SEPARATION OF THE GROUP IV CATIONS

The cations of Group IV all form soluble chlorides, sulfides, and hydroxides and are thus not precipitated by any of the reagents used to precipitate the other four groups.

Table 15.1
Solubility Properties of the Group IV Cations

	Ba^{2+}	Ca^{2+}
Cl^-	S	S
$\overset{.}{O}H^-$	S	S^-
SO_4^{2-}	I (white)	S^-
CrO_4^{2-}	A (yellow)	S
CO_3^{2-}, PO_4^{3-}	A (white)	A (white)
S^{2-}	S	S
Complexes	—	—

Note: S means soluble in water, >0.1 mole/l. S^- means slightly soluble in water, ~0.01 mole/l. A means soluble in acid ($6\,M$ HCl or other nonprecipitating, nonoxidizing acid). I means insoluble in any common solvent.

GENERAL SCHEME OF ANALYSIS

The procedure used for the qualitative analysis of Group IV is outlined in the flow chart given in Figure 15.1, and the detailed directions are given in Table 15.2.

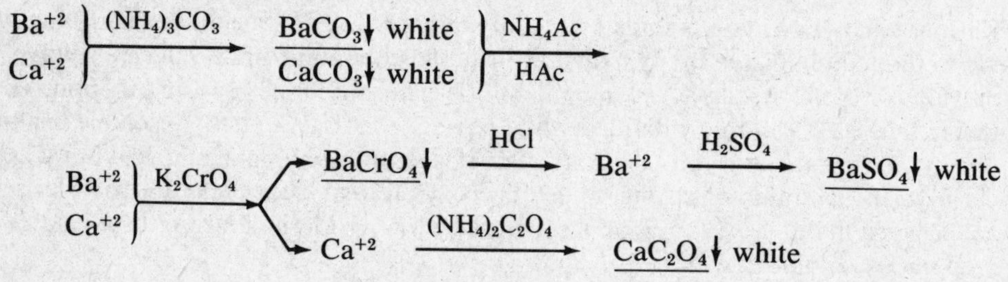

Fig. 15.1 *Group IV flow chart.*

Table 15.2
Group IV Flow Chart

Transfer the solution obtained from previous group separations to an evaporating dish and evaporate to dryness. Cool, then dissolve the residue in a mixture of 2 drops of 12 M HCl and 10 drops of H_2O. Make the solution basic with 6 M NH_3 solution and then add 2 drops in excess. Add 4 drops of 1 M $(NH_4)_2CO_3$, and heat in a bath for 3 minutes. Check for complete precipitation. Centrifuge and decant. Discard the decantate.

Precipitate: CaCO₃, BaCO₃ Dissolve the residue in 6 drops of NH₄Ac and 4 drops of 6 M HAc. Add 2 drops of 1 M K₂CrO₄. Centrifuge and decant.	**Solution: Ca⁺²** Add 6 M NH_3 solution until the color of the solution changes from orange to yellow. Add 4 drops of 0.2 M $(NH_4)_2C_2O_4$. A white precipitate of CaC_2O_4 confirms Ca⁺².

Precipitate: BaCrO₄
Wash with 8 drops of H_2O and discard wash liquid. Dissolve the residue in 2 drops of 12 M HCl and heat to boiling.

Solution: Ba⁺⁺
Add 2 drops of 6 M H_2SO_4. A white precipitate of $BaSO_4$ confirms Ba⁺².

DATA ANALYSIS

Construct a flow chart showing the behavior of your unknowns when subjected to analysis.

NAME   DATE

SECTION

SELF-STUDY QUESTIONS

1. Write balanced ionic equations for the following processes.
a. the separation of Ba^{2+} from Ca^{2+};

b. the confirmatory test for calcium;

c. the confirmatory test for barium.

2. A student made the following observations on his unknown cation mixture. Indicate the items necessary to complete the statements.

A sample of the unknown was treated with _____ to precipitate the first group; no _____ occurred. The solution was then treated with _____ and boiled, after which a dark precipitate formed. This precipitate was separated and the decantate saved for later experiments. The precipitate dissolved completely in nitric acid and the clear solution that formed was made basic with aqueous _____ to give a white precipitate which has the formula _____ and a colorless solution. Treatment of this colorless solution with _____ and heating in a water bath for three minutes yielded a yellow precipitate of _____.

The decantate from the group precipitation described above in the previous paragraph was made basic with NH_3, reagent _____ was added, and the mixture heated; a dark precipitate formed. The precipitate was separated and the decantate saved for future experimentation. This precipitate was completely soluble in HCl, which solution upon treatment with NaOH and H_2O_2 yielded a clear yellow-orange color. The yellow-orange color is due to the ion _____.

The decantate from the previous paragraph was treated with $(NH_4)_2CO_3$ to yield a white precipitate.

3. From the observations in problem 2 we can say that the cations _____ are definitely present and the cations _____ are definitely absent. The information is not sufficient to distinguish whether _____ are (is) present.

4. Describe the minimum number of steps required to identify the following:
a. Ca^{2+} in the presence of Ag^+;
b. Ba^{2+} in the presence of Fe^{3+};
c. Ca^{2+} in the presence of Ba^{2+};
d. Pb^{2+} in the presence of Ca^{2+}.

SECTION D

Gasimetric Analysis

EXPERIMENT

16

Analysis of a Nitrite Coordination Compound Using a Gas Evolution Method

I. INTRODUCTION

This experiment involves the analysis of a nitrite-containing compound by measuring the quantity of nitrogen it evolves in a chemical reaction. The analysis can be carried out on an unknown substance supplied by your instructor, or the method can be used to determine the purity of the coordination compound prepared in Experiment 10.

II. PROCEDURE

The analysis is performed with the apparatus shown in Figure 16.1.

Accurately weigh out two samples of your product into small test tubes; the weights should be about 0.1–0.15 g. Dissolve 0.5 g of sulfamic (*not* sulfuric) acid in

155

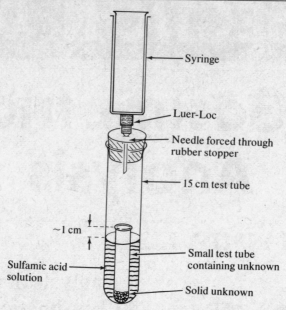

Fig. 16.1 *Apparatus for measurement of volume of nitrogen gas evolved from nitrite compound.*

10 ml of water. Carefully place one test tube containing the unknown into a 6 in. test tube and maintain it in an upright position. Carefully pipette the sulfamic acid solution into the large test tube (*not* into the small test tube), so that the level of the solution is about 1 cm below the edge of the small test tube. Do not get any of the sulfamic acid solution into the small test tube; if you do, the analysis will not give proper results. Gently, but firmly, insert the rubber stopper carrying the syringe needle into the 6 in. test tube. At this point the syringe should not be attached to the needle. After you are certain the rubber stopper is firmly in place, attach the syringe to the Luer-Lok fitting; record the reading on the syringe. The barrel of the syringe should be *lightly* greased so it moves easily under the pressure of the gas evolved. Mix the solid and the sulfamic acid solution by tilting the apparatus until the solution overflows the smaller test tube. The solid should dissolve and gas will be liberated from the mixture. Sulfamic acid reacts with nitrite ion according to Equation 1.

$$NO_2^- + NH_2SO_3H \longrightarrow N_2 + HSO_4^- + H_2O \qquad (1)$$

Be certain you get sufficient solution into the small test tube to have a complete reaction. Wait approximately 10 minutes and record (1) the final volume reading of the syringe, (2) the temperature of the room, and (3) the barometric pressure.

Repeat the analysis using the second sample. Calculate the percent nitrogen in your compound and the standard deviation.

III. DATA ANALYSIS

The weight of nitrogen liberated by your sample can be calculated from the volume of gas produced using the ideal gas law

$$PV = nRT \tag{2}$$

where P is the pressure of the gas, V is its volume, T is the absolute temperature, and n is the number of moles of gas present. R is the gas constant and its value depends upon the units used to measure the other quantities in Equation 2; some typically useful values of the gas constant are given in Table 16.1. The value of R in any other units can be obtained by using the appropriate conversion factors.

Table 16.1

Numerical value of R	Units
8.3143×10^7	ergs/deg mole
0.082054	liter-atm/deg mole
1.98726	cal/deg mole

The number of moles of a gas is given by

$$n = \frac{\text{wt}}{\text{MW}} \tag{3}$$

and substituting Equation 3 into Equation 2 we can obtain

$$PV = \frac{\text{wt}}{\text{MW}}RT \tag{4}$$

which upon rearrangement gives

$$\text{wt} = \frac{PV}{RT} \cdot \text{MW}. \tag{5}$$

In other words, the weight of a sample of nitrogen (N_2) can be easily calculated from a knowledge of the pressure, temperature, and volume of the sample. Thus, the percentage of nitrogen in a pure substance which weighs w_s is given in the usual way.

$$\text{percent N} = \frac{\text{wt of N}}{w_s} \times 100 \tag{6}$$

Substituting Equation 5 into Equation 6 gives

$$\text{percent N} = \frac{PV}{RT} \times \frac{\text{MW}}{w_s} \times 100. \tag{7}$$

For example, your experiment may yield the following results: 5.1346 g of a nitrite complex yielded 42.1 ml of N_2 measured at 25°C and a barometric pressure of 752 torr. From this information you should be able to determine the percentage nitrogen in this nitrite complex. The following analysis of the factors leading up to Equation 2 obtains:

  a. volume of gas = 42.1 ml or 0.0421 l
  b. pressure of gas: using Dalton's Law of partial pressures we can write

$$\text{atmospheric pressure } P_{total} = P_{N_2} + P_{H_2O}. \tag{8}$$

Since the nitrogen was collected over an aqueous solution we would expect the total pressure to consist of the pressure due to the N_2 generated and the vapor of water as generated by the solution in the flask. Thus, the pressure of N_2 is given by rearranging Equation 8.

$$P_{N_2} = P_{total} - P_{H_2O} \tag{9}$$

Strictly speaking, we should use the vapor pressure of the solution in the flask, but since we do not know the molar concentration of the salts dissolved in this solution, it is not possible to calculate the appropriate vapor pressure. We shall, as an approximation, use the vapor pressure of pure water, which can be looked up in Table 5.1, p. 43. At 25°C the vapor pressure of water is 23.7 torr. Thus the pressure of nitrogen is given as

$$P_{N_2} = 752 - 23.7 = 728$$

or

$$728 \text{ torr} \times 1 \text{ atm}/760 \text{ torr} = 0.958 \text{ atm}.$$

We calculate the pressure in atmospheres because we shall use the gas constant (Table 16.1) expressed in these units.

  c. Temperature:

$$25°C = (25 + 273)°k = 298°k$$

  d. Gas constant: we chose to use the gas constant (Table 16.1) expressed in liters, atmospheres, and °k; *viz.,*

$$R = 0.082054 \text{ l-atm/deg mole}$$

Substituting these quantities together with the weight of the sample and the fact that N_2 has a molecular weight of 28 g/mole, we obtain

$$\text{percent N} = \frac{(0.958)(0.042)}{(0.082054)(298)} \times \frac{(28)}{(5.1346)} \times 100 = 0.88 \text{ percent}$$

IV. ERROR ANALYSIS

Estimate the maximum uncertainty in the nitrogen analysis that arises from the following sources:

1. an error of $\pm 5°C$ is made in the temperature at which the volume of the N_2 is determined;

2. the volume of the N_2 is determined at 25°C, but a correction for the vapor pressure of water is not made;

3. the barometric pressure is in error by ± 5 torr;

4. the weight of the sample taken for analysis is in error by 0.001 g.

NAME   DATE

SECTION

SELF-STUDY QUESTIONS

1. Give the balanced half-reaction for the production of nitrogen from nitrite ion in acid solution.

2. Give a balanced half-reaction for sulfamic acid acting as a reducing agent.

3. Assume 0.4378 g of a nitrite complex yielded 45.7 ml of N_2 measured at 20°C and a barometric pressure of 741 torr.
   a. How many moles of N_2 were formed?

b. How many moles of NO_2^- must be present in the complex?

c. What is the percentage of nitrogen in the complex?

4. Describe the expected results of an analysis in a nitrite complex if sulfuric acid is used instead of sulfamic acid.

Volumetric Analysis

EXPERIMENT

17

The Determination
of the Acidity of a
Solution of Vinegar:
An Introduction to
Acid-Base Titrations

I. INTRODUCTION

One of the most important fundamental reactions in chemistry involves the reaction of an acid and a base:

$$\underset{\text{(acid)}}{HA} + \underset{\text{(base)}}{B} \longrightarrow A^- + HB^+ \qquad (1)$$

So long as either of the above reactants is a strong electrolyte—totally ionized—the reaction is quantitative (i.e., goes virtually to completion). In this experiment a strong base (NaOH) will be used to determine the concentration of acetic acid in an aqueous solution (vinegar). The reaction is given in Equation 2.

$$\underset{\text{OH}}{\overset{\overset{\displaystyle O}{\|}}{H_3C-C}} + OH^- \rightarrow \underset{\text{O}^-}{\overset{\overset{\displaystyle O}{\|}}{H_3C-C}} + H_2O \qquad (2)$$

The basic approach of *titrimetric analysis* is to deliver a known amount of a standard solution (the *titrant*) to a solution containing an unknown amount of a species which reacts in a known way with the titrant—in this case, according to Equation 2. The point at which the unknown species has been completely consumed is known as the *end point* of the titration; the reaction of an acid with a base (Eq. 2) is sometimes called a neutralization reaction.*

A. Indicator

The end point is found by the use of an indicator, a molecule that undergoes a characteristic chemical change (usually indicated by a color change) when a slight excess of the titrant solution is added. Indicators themselves are often molecules with acid-base properties, e.g., they undergo a process similar to that shown in Equation 3. Usually the

$$HIn \rightleftharpoons H^+ + In^- \tag{3}$$
$$\text{color 1} \qquad\qquad \text{color 2}$$

molecular form of the indicator (HIn) is a different color than its ionized form In$^-$. When the equilibrium shown in Equation 3 is disturbed (for example, by removal of H$^+$ from the solution when excess base is present after the end point is reached), the color of the solution changes from color 1 to color 2. In this experiment the indicator to be used is phenolphthalein, which is colorless in acid solution and red or pink in basic solution. The chemical change that occurs in this particular indicator is given by Equation 4.

The acid protons are indicated by an asterisk. The red form of phenolphthalein has a structure that is similar to that of many ordinary dyes. In effect, when a mixture of an acid such as acetic acid, and phenolphthalein is titrated with base, the acid reacts first. After the end point of the reaction is reached, addition of more base

---

*In general the end point is not at the point of exact neutralization where [H$^+$] = [OH$^-$]. The difference is small for the case discussed in this experiment, however.

causes reaction 4 to occur; the excess base reacts with phenolphthalein converting it into the red anion and the solution turns pink.

B. Standard Solutions

One of the difficulties in using a solution of sodium hydroxide as a titrant solution is that it is difficult to make up a standard solution of NaOH by direct weighing because NaOH is *hygroscopic* (i.e., it absorbs water from the atmosphere, thereby gaining weight during the weighing process). In such cases, the usual procedure is to make up a solution of NaOH of the approximate desired concentration and then to determine the concentration of the NaOH precisely by titrating it against a *primary standard*. In this experiment the primary standard is potassium acid phthalate (I, abbreviated as KHP):

(I)

(once again the acidic hydrogen atom is marked by an asterisk). This compound is non-hydroscopic, stable on drying, and a standard solution of KHP may be made up conveniently by weighing, using the usual techniques. Thus the present experiment is composed of two sets of titrations: (1) the preparation and standardization of a NaOH solution against KHP and (2) titration of unknown vinegar solution with the standardized NaOH solution.

NOTE: The accuracy of the present determination is equally dependent on both titrations.

II. PROCEDURE

A. Preparation of Sodium Hydroxide Solution

1. Obtain about 15 ml of 6 M NaOH solution. If any precipitate is present in the sample it should be removed by centrifuging.

2. Measure out about 13 ml of this solution in a graduated test tube and add to about 1 liter of distilled water in a bottle. Mix thoroughly by shaking. Label the bottle, leaving room on the label to add the correct normality after it has been determined in the next step.

B. Standardization of the NaOH Solution

1. Weigh out by difference (see Appendix 5) on a triple beam balance in a 30 ml beaker approximately 3 g of KHP. Dry this sample in an oven for at least an hour and cool in a desiccator.

2. Weigh out by difference 3 samples of dried KHP in three 250 ml Erlenmeyer flasks. Each sample should be approximately 0.8 g. Add approximately 75 ml of distilled water to each sample, and 2–4 drops of phenolphthalein indicator.

NOTE: The end point for this titration is described as follows. The first faint pink that persists in the solution for ~30 seconds or longer after swirling.

3. Rinse and fill a *clean* buret with the previously prepared NaOH solution. Titrate each sample of KHP to the end point and note the number of ml's of titrant used for each titration in your notebook.

4. Calculate the molarity (*M*) of the NaOH solution. Note the normality in your notebook and write it on the label of the NaOH solution.

C. Determination of the Acidity of the Unknown Vinegar Solution

1. Rinse a 5 ml pipette with small portions of the unknown vinegar solution. Then transfer 5 ml aliquots to three Erlenmeyer flasks. Add about 75 ml of distilled water and 2–4 drops of the phenolphthalein indicator.

2. Titrate the vinegar solution to the end point for all three samples. Note the volume of titrant used for each titration in your notebook.

III. DATA ANALYSIS

A. Calculation of NaOH Molarity

Suppose a sample of 0.8315 g of KHP required 31.10 ml of the NaOH solution to reach the end point. The molecular weight of KHP is 204.22 g. Therefore if *M* = molarity of the NaOH solution

$$M(0.0311) = \frac{.8315}{204.22}$$

therefore,

$$M = \frac{(.8315)}{(204.22).0311} = .1309.$$

Notice that the volume is expressed in liters.

The molarity for each titration should yield the same value of the normality to within approximately 2 percent.

B. Calculation of Acetic Acid Concentration

Suppose that the titration of a 5 ml aliquot required 15.6 ml of the standardized NaOH solution at the end point of the titration. The number of moles of acid must be the same as the number of moles of base (Eq. 5).

$$\text{moles acid} = \text{moles base} \tag{5}$$

Since the molarity of a solution is the ratio of the number of moles per liter, the number of moles of acid (or base) in a given volume, V (liters), is given by Equation 6 and Equation 7.

$$\text{moles acid} = V_a \times M_a \tag{6}$$
$$\text{moles base} = V_b \times M_b \tag{7}$$

Substituting Equation 6 and Equation 7 into Equation 5 gives

$$V_a M_a = V_b M_b \tag{8}$$

For the problem at hand Equation 8 becomes

$$M_a(0.00500) = M_b(0.0156) \tag{9}$$

Using the molarity calculated above in section III A for our hypothetic example (assuming that the average M was also 0.1309) Equation 9 becomes

$$M_a = \frac{(0.1309)(0.0156)}{(0.00500)} = 0.408 \tag{10}$$

The value of M_a for each titration should be reported, as well as the average for all three titrations and the standard deviation (Appendix 1). Approximately 2 percent reproducibility in the three M_a values may be expected with these techniques.

IV. ERROR ANALYSIS

Calculate the maximum uncertainty in the concentration of acetic acid that arises from the following sources:

1. There is an error of ± 0.02 ml in a buret reading during the standardization of NaOH solution.

2. The end point of the standardization of the NaOH solution is overshot by 0.5 ml.

3. The molarity of the NaOH solution is in error by ± 0.001 moles/l.

4. There is an error of ± 0.02 ml in a buret reading during the titration of the unknown acid.

5. The volume of the 5 ml pipette used to transfer the unknown acetic acid is in error by ± 0.1 ml.

NAME _____   **DATE** _____

SECTION _____

SELF-STUDY QUESTIONS

1. Give a balanced equation which represents the fundamental process that occurs in *all* acid-base reactions in aqueous solution.

2. Using LeChatelier's principle, describe the action of an indicator in a solution as the pH changes from acidic to neutral to basic.

3. Why is solid NaOH a poor primary standard?

4. Describe the characteristics of a good primary standard.

5. Assume that 0.7316 g of KHP required 28.64 ml of NaOH solution to neutralize it to a phenolphthalein end point.

   a. How many moles of KHP are present in the sample?

   b. What is the molarity of the NaOH solution?

6. Assume a 5.00 ml aliquot of acetic acid requires 14.7 ml of 0.1275 M sodium hydroxide solution.

   a. How many moles of acetic acid are present in the aliquot?

b. What is the molar concentration of acetic acid in the unknown solution?

EXPERIMENT

18

The Determination of Iron in a Sample: An Introduction to Redox Titrations

I. INTRODUCTION

A redox titration involves the same features as any titration: a *titrant* solution (the standard solution delivered from the buret), the unknown solution, and an *indicator* which undergoes a characteristic chemical change (usually accompanied by a color change) when there is a slight excess of the titrant in the solution. A feature of redox titrations that differs from the usual acid-base titration is that electrons are transferred between the titrant molecule and the species in the unknown solution. As an example, suppose that the titrant (Ox) is an oxidizing agent, so that the half reaction will be

$$ne^- + Ox \longrightarrow Ox^{-n} \tag{1}$$

while the species being titrated (Re) undergoes the half reaction

$$Re \longrightarrow Re^{+m} + me^-. \tag{2}$$

There is also present an indicator (Ind) which can be oxidized by the titrant but is less readily oxidized than the species Re. Thus when the titrant is initially added to the unknown solution the reaction that takes place is

$$mOx + nRe \longrightarrow mOx^{-n} + nRe^{+m} \tag{3}$$

(m)(n) electrons transferred.

When all of the Re molecules have reacted, then a small excess of Ox causes the reaction (4) to occur:

$$pOx + nInd \longrightarrow pOx^{-n} + n(Ind)^{+p} \tag{4}$$

(p)(n) electrons transferred.

The particular species used in this experiment are:
   a. potassium dichromate ($K_2Cr_2O_7$)—the titrant (an oxidizing agent);

b. ferrous iron (Fe^{+2})—the species titrated;
c. sodium diphenylamine sulfonate indicator.

A. Potassium Dichromate

Potassium dichromate is a very desirable oxidizing agent to use as a titrant because (1) it is an acceptable primary standard, i.e., a standard solution can be prepared by direct weighing, (2) it is stable, and (3) there are only two oxidation states (Cr^{+6} and Cr^{+3}) to consider. The half reaction of $Cr_2O_7^{-2}$ in acid solution is given by Equation 5.

$$Cr_2O_7^{-2} + 14H^+ + 6e^- \longrightarrow 2Cr^{+3} + 7H_2O \ (E^0 = +1.33v) \tag{5}$$

The chemistry and structure of the dichromate ion is relatively complex. For example, $Cr_2O_7^{-2}$ is in equilibrium with the chromate ion, CrO_4^{-2} (Eq. 6); the latter ion is used in the qual scheme (see Experiment 7) to precipitate Pb^{+2} or as a test for Cr itself.

$$\tag{6}$$

(tetrahedral, yellow)   (orange)

For this experiment, equilibrium 6 is shifted completely to the right by the addition of acid.

B. Ferrous and Ferric Ion

The unknown for this experiment is a sample of iron ore which in general will produce both Fe^{+2} (ferrous) and Fe^{+3} (ferric) ions upon dissolution in HCl. In order for the redox titration to be accurate *all* the iron in the sample must be reduced to Fe^{+2}. This is accomplished by the addition of $SnCl_2$ to a hot solution of Fe^{+3}. The reaction is

$$Sn^{+2} + 2Fe^{+3} \longrightarrow Sn^{+4} + 2Fe^{+2}. \tag{7}$$
2 electrons transferred

When iron is reduced from Fe^{+3} to Fe^{+2}, the solution changes color from light yellow (due to $Fe(H_2O)_6^{+3}$) to colorless or pale green (due to $Fe(H_2O)_6^{+2}$). Because Fe^{+2} can be easily oxidized by atmospheric O_2 as given by Equation 8,

$$2Fe^{+2} + \tfrac{1}{2}O_2 + 2H^+ \longrightarrow 2Fe^{+3} + H_2O \tag{8}$$

the ferrous ion solution must not be allowed to stand for a significant period of time before starting the titration.

A further complication exists. When reducing Fe^{+3} to Fe^{+2} with $SnCl_2$, a slight excess of the latter must be added. However, this excess would also react with the $Cr_2O_7^{-2}$ when the titration is started and more titrant would be consumed than there is iron in the sample. The excess Sn^{+2} is removed by adding $HgCl_2$ (mercuric chloride) which oxidizes Sn^{+2} (but *not* Fe^{+2}) as follows.

$$2HgCl_2 + Sn^{+2} \longrightarrow \underset{\text{(precipitate)}}{Hg_2Cl_2} + Sn^{+4} \tag{9}$$

In principle the Hg^{+1} ion could be oxidized by the $Cr_2O_7^{-2}$, but the fact that Hg_2Cl_2 is insoluble makes this reaction so slow that it does not interfere with the titration. Care must be exercised in adding the $HgCl_2$ so that the Hg_2Cl_2 is not further reduced by Sn^{+2} to elemental Hg, which would be readily oxidized by $Cr_2O_7^{-2}$ and would therefore interfere in the titration.

C. The Indicator

The indicator for this reaction is sodium diphenylamine sulfonate (I),

(I)

The chemistry this molecule undergoes when it is oxidized is complex but understandable. The first step (Eq. 10) is the removal of an electron and a hydrogen ion from (I), leaving a *free radical* (a molecule with an unpaired electron)

$$+ \ H^+ \ + \ e^- \ \text{(transferred to } Cr_2O_7^{-2}) \tag{10}$$

This step is followed by the dimerization of two free radicals at the point from which a hydrogen was removed. The structure of this product is given in (II).

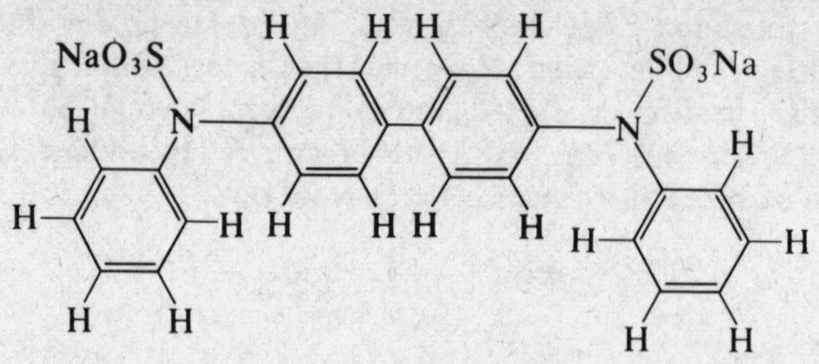

diphenylbenzidine sulfonate (colorless)

(II)

II is further oxidized (an electron is removed from each N atom), forming a long conjugated molecule, the structure of which (III) is

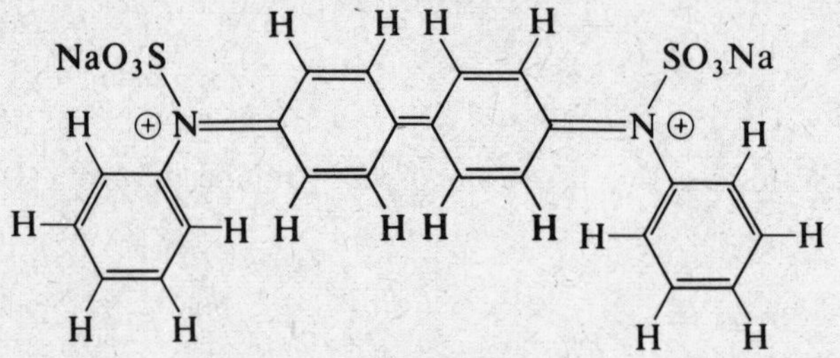

(violet) (III)

similar to that found in many dyes. It is the color of this molecule that serves as the end point of the titration.

II. PROCEDURE

This procedure is divided into three parts, (A) preparation of a standard 0.1N $K_2Cr_2O_7$ solution, (B) dissolution of iron ore sample, and (C) reduction-titration step.

NOTE: It is assumed that the student is familiar with the basic techniques of titration.

A. Preparation of Standard Solution of $K_2Cr_2O_7$

1. Weigh, on a triple beam balance in a 30 ml beaker 2.5 g of $K_2Cr_2O_7$. Dry the sample in an oven for at least 2 hours; cool in a desiccator.

2. Weigh by difference (see Appendix 5) the dry, *cool* sample accurately on the analytical balance into a 250 ml beaker. Add 100 ml of distilled water to dissolve the sample; carefully add this solution to a 500 ml volumetric flask and dilute to

the mark. If a 250 ml volumetric flask is used, then the sample of $K_2Cr_2O_7$ weighed out in (1) should be half as large.

3. Calculate the normality and enter the value in your notebook.

B. Dissolution of Sample

1. Obtain a sample of unknown. Weigh approximately 2.5 g on the triple beam balance, dry in an oven for at least 2 hours, and cool in a desiccator.

2. After the sample has cooled, weigh out three samples in the range of 0.5 to 0.7 g into three 500 ml erlenmeyer flasks. Weigh by difference using an analytical balance.

3. Add 10 ml of concentrated HCl and 10 ml distilled water to each sample. Heat solutions on the hot plate under the hood. Dissolution may require 30–60 minutes. If after 30 minutes of heating the mixture is cloudy, add 10 ml more of 12 M HCl and continue heating. The solutions may be safely stored between periods at this point. The total volume of each sample should be about 20 ml. Occasionally a white residue of silica remains after most of the ore has dissolved. This silica will not interfere with the titration and may be ignored.

C. Reduction and Titration

NOTE: Each ore sample, prepared as above, will be taken through the following procedure completely before the next sample is reduced and titrated.

Clean and fill the buret with the standard dichromate solution before beginning this procedure in order to eliminate any lengthy delay between reduction and titration.

1. If the solution is colorless add a few crystals of potassium permanganate ($KMnO_4$) which will oxidize some of the Fe^{+2} to Fe^{+3}. It is important that the solution be yellow (indicating that some Fe^{+3} is present) before adding the $SnCl_2$ so that the color change (yellow) $\longrightarrow$ (colorless) will indicate when sufficient $SnCl_2$ has been added.

> ☞ **CAUTION:** A large excess of $SnCl_2$ is to be avoided.

2. Before addition of the $SnCl_2$ the solution should be heated (but not boiled). Add the $SnCl_2$ dropwise. After the yellow to colorless color change has taken place, add one or two drops more of $SnCl_2$.

3. Cool the solution under the tap and add 20 ml of saturated mercuric chloride ($HgCl_2$) all at once to the sample. A white precipitate of Hg_2Cl_2 will result. If the precipitate is gray or black some free Hg has been formed and the sample should be discarded. If no precipitate forms after a few moments the sample also will have to be discarded.

4. Allow the sample to stand for a few minutes. Add 175 ml distilled water, 5 ml H_2SO_4 (concentrated) and 5 ml of 85 percent H_3PO_4, and 8 drops of the diphenylamine sulfonate indicator.

5. Begin to titrate. The solution will become green from the formation of Cr^{+3} ions. The end point is indicated by a violet or purple color. The end point of this titration is very sharp, the addition of even half a drop of titrant at the end point will cause a dramatic color change from dirty green to deep purple. Go slowly near the end point.

6. Enter the amount of titrant to reach the end point of the titration in your notebook.

This process is repeated for the other two samples. It may be advisable to refill the buret at this point so that the second titration will not have to be interrupted.

III. DATA ANALYSIS

Calculate the weight percent of iron in the sample for each sample. The equivalent weight of Fe is 55.85.

Suppose 2.304 g of $K_2Cr_2O_7$ were dissolved in 500.0 ml water. The normality of the solution would be

$$N = \frac{\text{weight of } K_2Cr_2O_7}{(\text{eq. wt. of } K_2Cr_2O_7)(V)} = \frac{2.304}{(49.04)(0.5000)} = 0.09396.$$

Note that V is expressed in liters.

The eq. wt. (equivalent weight) is the number of grams/electron transferred. Since six electrons are transferred to the $Cr_2O_7^=$ ion,

$$\text{eq. wt. } K_2Cr_2O_7 = \frac{\text{formula weight } K_2Cr_2O_7}{\text{number of electrons transferred}} = \frac{294.21 \text{ g}}{6 \text{ electrons}}$$

$$= \frac{49.04 \text{ g}}{\text{electron}}$$

The number of equivalents* of titrant used during a titration is equal to NV_t where N is the normality of the titrant solution and V_t is the volume in *liters* of titrant used. The number of equivalents of titrant equals the number of equivalents of the species being titrated. The number of grams of unknown is given by

grams of unknown = (number of equivalents of unknown) (eq. wt. of unknown).

For Fe^{+2} the equivalent weight equals the formula weight (55.85) because one electron is transferred to produce Fe^{+3}, therefore

$$\text{g of Fe in sample} = NV_t(55.85).$$

Suppose $V_t = 27.8$ ml, then

$$\text{g of Fe} = (0.09396)(0.0278)(55.85) = 0.1460.$$

Suppose the original sample of iron ore weighed 0.6342 g. The weight percentage

---

*An equivalent is analogous to a mole. One equivalent is the number of grams of a substance required to produce one mole of electrons in a redox reaction.

is given by

$$\frac{\text{g of Fe}}{\text{weight of sample}} \times 100 = \frac{0.1460}{0.6342} \times 100 = 23.02 \text{ percent.}$$

Report the average value for three determinations and the standard deviation (Appendix 1).

IV. ERROR ANALYSIS

Calculate the maximum uncertainty in this analysis that arises from the following sources.

1. An error of ± 0.001 g is made in the weight of $K_2Cr_2O_7$ used to prepare the standard oxidizing solution.

2. An error of ± 1 ml is made in the total volume of the standard solution when it is prepared.

3. An error of ± 0.02 ml is made in the volume of the standard solution used to titrate an iron ore sample.

4. The end point of the titration is overshot by 1.0 ml.

Describe the nature of the error expected in this analysis if:

5. the titration is carried out on the iron ore sample immediately after it is dissolved in HCl.

6. the titration with $K_2Cr_2O_7$ is carried out after the dissolved ore is treated with $SnCl_2$ but before $HgCl_2$ is added.

7. the $K_2Cr_2O_7$ is not dried before it is weighed in the course of making up the standard solution.

8. the iron ore sample is dissolved and the iron gotten into Fe^{2+} form, but the titration is carried out two days later.

NAME _____   DATE _____

SECTION _____

SELF-STUDY QUESTIONS

1. Give a balanced half reaction for $K_2Cr_2O_7$ acting as an oxidizing agent.

2. Describe, using equations where appropriate, the process whereby all the iron in the unknown sample is gotten into the 2+ oxidation state prior to titration with $K_2Cr_2O_7$.

3. Assume 2.6301 g of dry $K_2Cr_2O_7$ is dissolved in 250 ml of solution.
a. How many moles of $K_2Cr_2O_7$ are present?

b. How many equivalents of $K_2Cr_2O_7$ are present?

c. What is the molarity of the solution?

d. What is the normality of the solution?

4. If 3.7171 g of iron ore, after proper treatment, require 35.65 ml of the solution described in problem 3 to reach an end point:
    a. how many equivalents of iron are present?

b. how many moles of iron are present?

c. how many grams of iron are present?

SECTION F

Gravimetric Analysis

EXPERIMENT

19

The Analysis of
Silica in Cement: A
Gravimetric
Determination

I. INTRODUCTION

Gravimetric analysis is one of the oldest and most useful analytical methods available to chemists. Its basis is the law of constant proportions, which states that the chemical composition of pure substances is fixed. Gravimetric analysis generally involves (a) obtaining the weight of the sample, (b) treating the sample chemically to obtain the component of interest in a form that can be weighed, (c) and weighing this product. Often the component of interest is precipitated from solution, but it is equally possible to remove all but the component of interest by leaching or by driving off volatile substances. In this experiment we shall determine the percent of silica (SiO_2) in cement by converting all other elements present into soluble substances.

A typical portland cement contains about 21 percent silica as well as compounds of calcium, magnesium, iron, aluminum, sodium, and potassium. Cement is manufactured by heating a mixture of limestone ($CaCO_3$), cement rock, clay, and shale; this product is then ground to a fine powder. Adding water to this product causes reactions to occur that produce a hard mass which principally consists of interlocking crystals of various calcium silicates and calcium aluminates.

The analysis described in this experiment consists of converting all components except SiO_2 into soluble substances; the silicon-containing species are converted into hydrated silica (Eq. 1) which is filtered on ashless filter

$$CaSiO_3 + 2HCl \longrightarrow CaCl_2 + SiO_2 \cdot H_2O \tag{1}$$

paper and ignited to 800°C to produce anhydrous SiO_2 (Eq. 2).

$$SiO_2 \cdot H_2O \xrightarrow{heat} SiO_2 + H_2O \tag{2}$$

II. PROCEDURE

A. Weighing Analytical Sample

Weigh about 2 g of cement into a clean 250 ml beaker using an analytical balance. The cement should be pulverized (in a mortar), and the weighing done by difference from a small beaker containing the pulverized cement. The cement should be thoroughly pulverized—otherwise the sample will not dissolve completely or rapidly.

B. Dissolution of Sample

Transfer about 2 g (exact weight is unimportant) of solid ammonium chloride to the beaker containing the cement. The ammonium chloride may be weighed with a triple beam balance on a piece of paper and transferred to the beaker. Add 15 ml of dilute ($6N$) hydrochloric acid to the beaker. Stir to dissolve the ammonium chloride. Put a rubber policeman on a stirring rod and stir the mixture with the *uncovered end* of the stirring rod. The policeman end will be used later. Using the hotplate in the hood, evaporate the solution without boiling; keep the stirring rod in the beaker while the solution is evaporating. Keep the solution from splattering; if the hot plate is too hot, place a wire gauze under the beaker. It is important that you make certain that all the cement reacts with HCl.

C. Preparation of Crucible

While evaporation is going on, prepare to heat a clean, dry crucible (Fig. 19.1). The crucible may need to be cleaned in hot aqua regia (hood) before you heat it. Heat the dry crucible gradually with the burner, then strongly (until the bottom glows red) for 30 seconds. Allow it to cool in place to room temperature.

Handling the crucible with tongs (this procedure will be demonstrated by your laboratory instructor), put it in a clean, dry evaporating dish and carry it to the

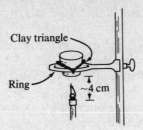

Fig. 19.1 *Apparatus heating a substance in a crucible. Heating is begun with a low flame. Note proper distance between burner and crucible.*

analytical balance. Weigh and record the weight. Return it to your bench. If you plan to do multiple experiments, mark the bottom of the crucibles with cobaltous chloride solution, which may be applied with a match stick. After application, heat the crucible in a flame; this process should leave a permanent identifying mark.

D. Filtration of Cement Mixture

When your digested cement mixture has dried out, bring the beaker to your bench. Remove the triangle from the ring and replace it with a screen. Pour about 100 ml of distilled water into a 400 ml beaker and put it on the screen. Heat to boiling.

While the water is heating, add 50 ml of water and 5 ml of dilute hydrochloric acid to the sample mixture. Stir.

Mount a funnel in a ring (Fig. 19.2). Fold a piece of ashless filter paper, as demonstrated, and put it in the funnel. Wet it with water (wash bottle) and secure its fit. Put a 250 ml flask below the funnel to catch the filtrate.

When the distilled water is near boiling remove the beaker (handle it around the rim) from the screen, and put the beaker with the sample mixture on the screen. Heat the mixture to redissolve soluble salts. Rinse the stirring rod in the beaker using your wash bottle; then remove the rod.

When the sample solution is hot, handle the beaker around the rim and carefully

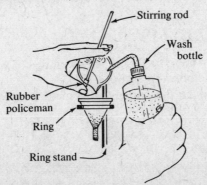

Fig. 19.2 *Rinsing a beaker after filtration. The rubber policeman is used to scrub beaker walls free of residue.*

pour part of the solution into the funnel. As it drains, add more until all liquid has been transferred.

As illustrated in Figure 19.2 and as demonstrated by the instructor, rinse the beaker with water from the wash bottle. Add about one-third of the hot water, scrub with a rubber policeman, and pour the washings into the funnel. Repeat until all hot water is used up. The residue ($SiO_2 \cdot H_2O$) should be washed thoroughly with hot distilled water until all pigmented impurities (which give the residue a yellow color) are removed.

E. "Ashing" the Filter Paper

After the last wash water has drained, gather the paper, as demonstrated, and place it in the weighed crucible. Replace the screen on the ring with the clay triangle and put the crucible on it. Apply heat gradually to dry the paper, then heat more strongly to char the paper, and finally, with an asbestos board under the burner, heat until the crucible bottom is bright red. Use a crucible lid after the paper has been charred to keep the contents from splattering while heating strongly. The lid should not be placed fully on the crucible; it should be slightly displaced.

When all of the black carbon particles have disappeared and a gray-white mass remains in the crucible, stop heating. Allow the crucible to cool in place.

When the crucible is at room temperature, handle it as before (tongs, evaporating dish) and obtain its weight. Record.

III. DATA ANALYSIS

The weight of the sample and the undissolved silica is obtained by difference.

The percent silica in your cement sample is obtained in the usual way (Eq. 3).

$$\text{percent silica} = \frac{\text{weight of unreacted residue}}{\text{weight of sample}} \times 100 \qquad (3)$$

IV. ERROR ANALYSIS

Determine the maximum uncertainty in the percentage of silica in a sample of cement that arises from the following sources.

1. The beaker into which the pulverized cement sample is weighed carries a film of water that weighs 0.010 g.

2. A portion of the reaction mixture containing the dissolving cement sample is lost by splattering.

3. The crucible in which the silica sample is to be ashed is coated with a film of water that weighs 0.010 g.

4. The hydrated silica is ashed while it is yellow colored.

5. The crucible containing the filter paper and hydrated silica is heated until black specks of carbon remain and then weighed.

6. The ashing is carried out for a short time with a tight lid on the crucible.

NAME   DATE

SECTION

SELF-STUDY QUESTIONS

1. What is the form of silicon that is weighed in this analysis?

2. How is anhydrous silica produced in this analysis?

3. Give a balanced equation for the conversion of calcium silicate ($CaSiO_3$) into hydrated silica.

4. Explain why it is important to have a dry crucible for the ashing process.

5. Describe how you know that the hydrated silica contains no impurities and is ready for ashing.

6. Give a balanced equation for the dehydration of the residue in the ashing process.

7. Explain why the crucible should *not* be tightly closed during the ashing process.

SECTION G

Liquids and Solutions

EXPERIMENT

20

Gravimetric
Determination of
Nickel with
Dimethylglyoxime

I. INTRODUCTION

This experiment is an example of *quantitative analysis,* which means that an accuracy of better than 1 percent can be expected. The greatest care must be taken with all weighings and other operations in the detailed procedure. In particular the same balance should be used for all weighings. Students should be familiar with the proper use of the analytical balance before beginning this experiment. Absolute accuracy on the order of ± 0.005 g or better is required for this experiment.

The overall chemical reaction that is to be carried out is shown in Equation 1.

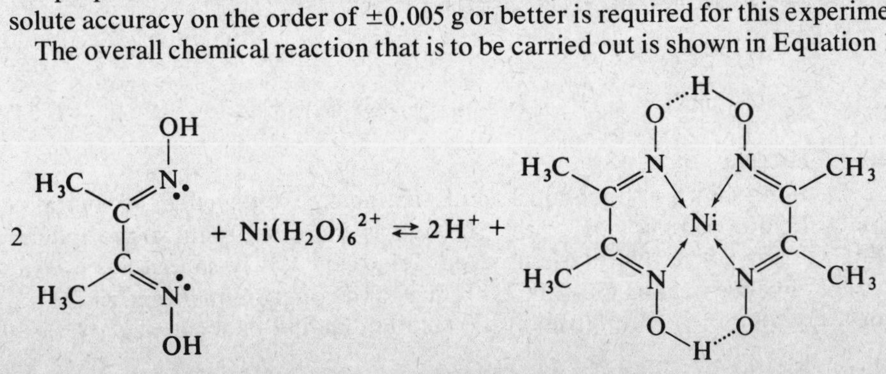

$$2 \quad \text{(dimethylglyoxime)} + Ni(H_2O)_6^{2+} \rightleftharpoons 2H^+ + \quad \text{(nickel complex)} \tag{1}$$

Nickel dimethylglyoxime is an example of a *chelate,* in which the unshared electrons of the nitrogen form a bond with the metal atoms. Note from the equation that the reaction produces H^+ ions, so the reaction can be driven to form products in a basic solution. This is why it is important to control the pH of the solutions used in this procedure.

One of the practical considerations of a gravimetric analysis of the present type is the nature of the precipitate. It is desirable to produce a precipitate with a particle size as large as possible because of the increased ease of filtration, the reduction in the tendency of the precipitate to "creep" and to stick to the glass walls, and a decrease in the total surface area of the precipitate. Since impurities tend to be absorbed on the surface of the precipitate, the decrease in surface area leads to a precipitate of greater purity.

If a precipitate begins to form *homogeneously* throughout the whole solution then relatively few *nucleation centers* are formed. Each nucleation center is a "seed" for a larger crystal, and if there are relatively few seeds then the precipitate will be composed of relatively few but also relatively large particles. By contrast, if the precipitate is formed upon mixing of two chemicals, many more nucleation centers will be formed along the surface of mixing of the two chemicals *(heterogeneous precipitation),* and the average particle size will be diminished. For this analysis the ease of handling of the precipitate is *very* dependent on particle size. Thus it is important that the appropriate homogeneous precipitation scheme be followed carefully.

The method employed in this experiment is to control the pH of the solution. The Ni^{+2} solution is diluted with a solution of HCl at $pH = 1.75$. No $Ni(DMG)_2$ (DMG = dimethylglyoxime) could form in this solution because it is too acidic. Urea is then added to the mixture and dissolved. Urea undergoes the *hydrolysis reaction* shown in Equation 2.

$$\begin{matrix} H_2N \\ \\ H_2N \end{matrix} \!\!\!\! C = O + 2H_2O = 2NH_4^+ + CO_3^{-2} \qquad (2)$$
(urea)

The carbonate ion formed in the hydrolysis reaction reacts with excess hydrogen ion in solution (Eq. 3 and Eq. 4) because it is the anion

$$CO_3^{-2} + H^+ = HCO_3^- \qquad (3)$$
$$HCO_3^- + H^+ = H_2CO_3 \longrightarrow H_2O + CO_2 \qquad (4)$$

of a weak acid (H_2CO_3).

The reactions shown in Equation 3 and Equation 4 have the effect of neutralizing the solution, and the pH of the solution is increased until a precipitate, $Ni(DMG)_2$, forms. The hydrolysis proceeds relatively slowly so that a *complete lab period* of gentle heating (50–60°C) is required to complete the reaction. At the end of the period or before filtration the solution should be made slightly basic with a few drops of NH_4OH. Also the solution should be checked for complete precipitation by the addition of a few drops of the DMG solution.

II. PROCEDURE

1. Carefully clean three 15 ml Gooch crucibles (using the vacuum flask to draw a few milliliters of concentrated HNO_3 through the frits followed by distilled H_2O is advised). Take great care to rinse well with distilled H_2O. Once the cleaning process has begun do not handle these crucibles with your fingers; always use the crucible tongs or a piece of paper to protect the glass surfaces from greasy fingers. Number the crucibles with a *pencil* on the ground glass surface. After cleaning, the crucibles should be dried for about an hour in an oven. It is convenient to place all Gooch crucibles in a large beaker while drying. After drying, the hot crucibles are removed to a desiccator to cool for 15–20 minutes before weighing. Weigh each crucible to the nearest 0.0001 g on the analytical balance and record in your notebook. Repeat this procedure. The first and second weight of the crucible should agree to ±0.0005 g. Store the crucibles in your desiccator.

> ☛ **CAUTION:** The accuracy of this whole analysis is as dependent on this step as any other step to follow. The weight of the precipitate ultimately will be determined by difference so an accurate crucible weight is of critical importance.

2. Pipette three 5 ml aliquots of the nickel unknown solution into three clean beakers (250 ml or 400 ml size). Dilute to a total volume of 100 ml with the stock solution of pH 1.75 HCl. Label each beaker with your name.
3. Weigh out three ~15 g (±1 g) samples of urea on a triple beam balance and set aside.

> ☛ **CAUTION:** For steps 4 and 5 treat the beakers one at a time, and reasonably quickly (30–60 sec/beaker).

4. Add 15 g of urea to each beaker and stir rapidly until dissolved.
5. Add ~30 ml of the stock 1 percent DMG solution to each beaker. Stir well for ~10 seconds. No precipitate should form. If it does, the homogeneous precipitation has been subverted (probably the pH of the stock solution is incorrect) and the experiment should not be continued.
6. Cover each beaker with a watch glass and place on the hot plate. The solution should never boil; a temperature of 50–60°C is ideal. A red precipitate should form after 2–10 minutes. If no precipitate forms after 30 minutes the experiment should be aborted (either the pH of stock HCl solution was too low or not enough urea was added).
7. The beakers should be heated until the pH is 7 or 8 (use your pH paper). More urea may be added after two hours of heating if the pH seems to be rising too

slowly. If the pH of the solution does not change after two laboratory periods of heating add a few drops of dilute NH_4OH. These solutions may be stored and the heating resumed at a later lab period.

8. Before beginning to filter, a few drops of DMG solution should be added to each solution to check for complete precipitation. If new precipitate is observed to form add ~5 ml more of DMG. Check the pH of the solution. If acid ($pH < 7$), add more urea and heat until the pH is 7 or 8. Before filtering add 1 or 2 drops of dilute NH_4OH to each solution. The contents of the beaker should be vacuum filtered using the Gooch filter crucibles previously weighed. Each filtration should require no more than 15–20 minutes, if you pay attention to your work. The crucible should never be filled more than two-thirds of the full volume.

☛ **CAUTION:** The appropriate technique for a vacuum filtration using a Gooch crucible should be known from either lecture demonstrations or the laboratory instructor.

---

9. Wash the precipitate with three 10 ml portions of distilled water to which a few drops of concentrated NH_4OH have been added. Place the three Gooch crucibles in a 400 ml beaker and dry for at least 2 hours in the oven.

10. After drying, remove the crucibles to the desiccator to cool for 20 minutes and weigh carefully. For the most careful work the crucibles should be redried and reweighed to ensure complete removal of water.

Note on time required for this analysis: The cleaning of the Gooch crucibles and the formation of the precipitate (steps 1–6) will require most of a lab period (3 hours). The Gooch crucibles should be dried for the period between lab periods. The filtration will require a significant fraction of a lab period. A proposed timetable is as follows:

period 1: clean crucibles (step 1), dry and weigh (15–20 minutes of operations).

period 2: carry out precipitation (steps 2–7) (full lab period).

period 3: filter precipitates (steps 8–9) and dry. Remove and place in desiccator at end of period.

period 4: finish drying precipitate (30–45 minutes of operations).

The above timetable allows other experiments to be carried out while the nickel analysis is going on.

III. DATA ANALYSIS

Suppose the empty dry Gooch crucible weighed 15.3427 g and weighed 15.5842 g, 15.5820 g, and 15.5818 g after filtering the precipitate and drying for 2 hours, 3 hours, and 4 hours respectively. The last 2 weighings are in reasonable agreement indicating that all the water has been removed from the precipitate. The last weighing is taken to be the correct weight of the completely dry crucible plus precipitate. The net weight of the precipitate is calculated as

$$15.5818 \text{ g crucible } + \text{ precipitate}$$
$$\underline{-15.3427 \text{ g crucible}}$$
$$0.2391 \text{ g precipitate.}$$

Next we need to calculate the number of mg of Ni contained in this precipitate. The formula weight of Ni(DMG)$_2$ is 288.94 and the atomic weight of Ni is 58.71. If we multiply the weight (in mg) of the precipitate (which is Ni(DMG)$_2$) $\times$ the ratio (58.71/288.94) we obtain the number of mg of Ni in the precipitate.

$$239.1 \times \frac{58.71}{288.94} = 48.48 \text{ mg of Ni}$$

This number of mg was contained in the original 5 ml aliquot of the unknown solution, so the concentration of Ni in the unknown solution is 48.55 mg/5 ml = 9.717 mg/ml.

You will obtain 3 independent values for this concentration. They can be expected to agree within ± 1 percent (i.e., $\pm 9.717 \times 0.01 = \pm 0.1$ mg/ml).

Report the three values of mg/ml determined by your analysis, the average value, and the standard deviation (Appendix 1). Points will be assigned on the basis of the accuracy of each independent determination. In this way one is not penalized so severely if one determination is very inaccurate.

IV. ERROR ANALYSIS

Calculate the maximum uncertainty in the analysis that arises from the following sources.

1. An error of ± 1 mg in the weight of the precipitate.

2. An error of ± 1 mg in the weight of any given weighing.

3. An error of ± 1 mg in each weight involved in a given determination.

4. The 5 ml pipette used to transfer the solution has a volume error of ± 0.1 ml.

5. Describe the type of error you would expect if the pH of the solution during the precipitation never exceeded 6.0.

6. What would be the error in the determination of nickel if the sintered glass disk of the crucible contained NaCl from a previous experiment and was not washed before use in this experiment?

NAME _____ DATE _____

SECTION _____

SELF-STUDY QUESTIONS

1. Give balanced equations for the processes which occur to change the pH of the solution in this experiment.

2. Write an equation which shows dimethylglyoxime acting as an acid in aqueous solution.

3. Describe the method used to ensure that a precipitate has been thoroughly dried.

4. Describe the advantage of using a Gooch crucible in gravimetric analysis compared with filter paper.

5. Assume a 5.00 ml sample of a nickel-containing aqueous solution yields 0.1741 g of $Ni(DMG)_2$.
   a. How many moles of nickel are present in the sample?

   b. What is the concentration, expressed in mg/ml, of nickel in this sample?

EXPERIMENT

21

Semi-quantitative Separation of a Two-Component Organic Mixture by Extraction

I. INTRODUCTION

The present experiment is concerned with separating a mixture of benzoic acid (I) and naphthalene (II) by the technique of extraction.

(I)   (II)

The mixture is dissolved in methylene chloride (CH_2Cl_2, b.p. 40.1°C) and shaken in a separatory funnel with an ~1 M NaOH aqueous solution. The benzoic acid is converted to the benzoate anion as indicated in Equation 1.

$$+ \quad OH^- \quad \rightarrow \qquad\qquad + \quad H_2O \tag{1}$$

Sodium benzoate is far more soluble in the aqueous layer than the organic layer (CH_2Cl_2) because the highly polar H_2O molecules efficiently solvate any charged species. As a result of reaction 1 there is a transfer of the benzoic acid (in the form of the anion) from the organic layer to the aqueous layer. Naphthalene is not affected by NaOH and is essentially insoluble in H_2O. Thus the result of this procedure is to transfer the benzoate anion into the aqueous layer while leaving the naphthalene in the organic layer. Since CH_2Cl_2 and H_2O are immiscible, the two layers are easily separated.

The benzoic acid is recovered from the benzoate anion by making the aqueous layer acid (i.e., ~0.2 M), reversing the neutralization reaction (Eq. 2).

$$(2)$$

The un-ionized benzoic acid is not very soluble in H_2O (0.18 g/100 ml at 4°C, 0.27 g/100 ml at 18°C) and precipitates out. The solution is cooled to 5–10°C in order to precipitate the benzoic acid as completely as possible before filtration.

The naphthalene is recovered by removing the organic solvent by evaporation, leaving behind the much less volatile naphthalene.

The object of this experiment is to carry out the steps described above to separate an unknown mixture of benzoic acid and naphthalene into its two components and to determine the percentage by weight of each component in the original mixture. While the objective for this experiment is the determination of the percentage composition by weight, you should calculate the percentage of the material recovered as a check on the reliability of results. In general the *yield*, given by expression 3,

$$\text{yield} = \frac{\begin{array}{c}\text{(weight of benzoic acid recovered)} \\ + \text{ (weight of naphthalene recovered)}\end{array}}{\text{(weight of starting mixture)}} \times 100 \quad (3)$$

should be in the range 85–115 percent. A yield over 100 percent implies that the solid material is still wet with solvent and should be air-dried further.

The present separation is not very accurate and so is classified as a semi-quantitative procedure. In a true quantitative analysis we may expect an accuracy of parts per thousand (i.e., 0.1 percent), while the present procedure is likely to yield errors of the order ±10 percent. While it is not required, it is useful to repeat these procedures at least once to check the reliability of your final result.

II. PROCEDURE

1. Weigh out approximately 3 g of the unknown mixture (accuracy of weighing should be ±0.01 g) on the analytic balance and transfer to a 250 ml erlenmeyer flask.

2. Dissolve the mixture in 25 ml of methylene chloride (CH_2Cl_2). If after a few minutes stirring all the unknown has not dissolved, add more CH_2Cl_2 a few milliliters at a time (with stirring) until all solid material has dissolved. The tota

volume of the organic solvent added should be less than 35 ml. Transfer the solution into a 250 ml separatory funnel.

3. Make an approximate 1 M NaOH solution by dissolving ~10 g of NaOH in 250 ml water. Alternatively 10 ml of 8 M NaOH stock solution can be diluted to 80 ml with distilled water using your graduated cylinder.

4. Extract the solution twice with 25 ml of the 1 M NaOH solution.

NOTE: The organic layer is the bottom layer.

Each time the organic layer is transferred into the previously used 250 ml erlenmeyer flask and the aqueous layer is transferred into a 250 ml beaker.

> ☛ **CAUTION:** Do not undertake this step until the proper use of a separatory funnel has been demonstrated. Be sure you understand the meaning of the instruction "extract the solution two times" in step 4 above.

5. Stopper the erlenmeyer flask containing the organic layer and put aside until you are ready to evaporate the organic solvent. Alternatively the erlenmeyer flask may be left unstoppered in your desk for a day or two (but no longer or a significant amount of naphthalene may sublime) in which case most or all of the CH_2Cl_2 will evaporate. Record the weight of naphthalene in your notebook.

6. Add approximately 10 ml of concentrated (12N) HCl to the aqueous solution. Flakes of benzoic acid will precipitate out. It is necessary for the solution to be quite acidic to force all possible benzoic acid out of solution. It is advisable to check the solution with your pH paper (a pH less than 2 is recommended). Cool the solution in an ice bath for about 15 minutes before filtering.

7. Vacuum filter the aqueous solution. Allow the benzoic acid to dry in the air or in your desiccator between lab periods before weighing. Record the weight of the benzoic acid in your notebook.

> ☛ **CAUTION:** Benzoic acid is quite light and flaky and is easily blown away. Be careful in storing and handling this material.

III. DATA ANALYSIS

Calculate the weight percentage of each component of the mixture using the relationship in Equation 4 and Equation 5:

$$\text{weight percent benzoic acid} = \frac{\text{weight benzoic acid recovered}}{\text{initial weight}} \times 100 \quad (4)$$

$$\text{weight percent naphthalene} = \frac{\text{weight naphthalene recovered}}{\text{initial weight}} \times 100 \quad (5)$$

where

initial weight = initial total weight of the unknown mixture.

As a check on the efficiency of the recovery calculate the yield by the expression given in Equation 6:

$$\text{yield} = \frac{\text{total weight recovered}}{\text{initial weight}} \times 100 \qquad (6)$$

IV. ERROR ANALYSIS

1. The solubility of benzoic acid is 0.18 g/100 ml at 4°C and 0.27 g/100 ml at 18°. What is the expected absolute and percentage error in the recovered benzoic acid for 3 g unknown sample that is 40 percent benzoic acid if the precipitation occurs at 4°C and 18°C?

2. If the organic layer is extracted with three 25 ml portions of 1 M NaOH instead of 2, there will be a more complete removal of benzoic acid from the organic layer. However, there will be a larger loss of benzoic acid from its finite solubility in water. How large would this additional loss be, as a percentage of the mass of naphthalene you actually recovered?

3. If possible, weigh the CH_2Cl_2-naphthalene sample each day for several days, well after the CH_2Cl_2 appears to be gone. Estimate the daily sublimation loss of the naphthalene and calculate the significance of this loss on your results.

4. Suppose instead of water a solvent was used in which naphthalene has a solubility of 0.2 g/100 ml at room temperature and 0.05 g/100 ml at the temperature of the precipitation. What error would this yield for the percentage of naphthalene and benzoic acid?

5. Considering various possible errors (such as the ones above) estimate the expected error in the percentage composition of the unknown.

NAME

DATE

SECTION

SELF-STUDY QUESTIONS

1. In order for an extraction to work, what must be true of the solvents involved and the solubility properties of the solute molecules?

2. Why does benzoic acid transfer to the 1 *M* NaOH aqueous layer? Why is naphthalene so insoluble in the aqueous layer?

3. Describe or sketch a separatory funnel and describe how an extraction works.

4. Why is concentrated HCl added to the 1 *M* NaOH solution containing the benzoate ion? Why can't the aqueous layer just be evaporated to recover the benzoic acid?

5. What is the meaning of a yield of greater than 100 percent in this experiment? If such a yield is obtained, what should you do?

6. Suppose your initial weight of unknown was 2.74 g and you obtained 1.36 g of benzoic acid and 1.62 g of naphthalene. Calculate the weight percent of each component and the yield (see question 5 above concerning the yield).

APPENDIX 21. THE USE OF THE SEPARATORY FUNNEL AND EXTRACTIONS

A drawing of a separatory funnel is presented in Figure 21.1. The peculiar shape of this piece of glassware is designed to facilitate the separation of immiscible liquids. When carrying out a separation the only things that can go wrong are that the experimenter fails to close the stopcock in time (or equally important, fails to allow all the lower layer of liquid to drain out completely) or that the stopcock leaks. The latter should be checked for carefully before using a particular separatory funnel in an experiment. Also, when draining out any liquid, remember to remove the ground glass stopper to ensure a smooth flow of liquid.

In carrying out an extraction the following steps are required:

1. Add all liquids to the separatory funnel, *after checking to ensure that the stopcock is closed!*

2. Replace the ground glass stopper, which must fit sufficiently tightly that liquid does not leak out, and while holding the stopper in place with one hand, invert the separatory funnel as shown in Figure 21.2, and vent the contents by opening the stopcock. *Be sure and close the stopcock after venting.* (This step is required because there may be heat liberated upon mixing, which is especially true of this experiment, and the resulting heat may cause an increase of pressure inside the separatory funnel; this pressure must not be allowed to build up or the stopper may be pushed out of place.)

3. Shake the contents of the separatory funnel fairly vigorously, causing good contact between the two immiscible layers (too vigorous shaking can cause an emulsion, so don't go overboard). Vent the separatory funnel every minute or so.

4. After a few minutes of shaking, return the separatory funnel to its normal upright position (Fig. 21.1), and let the layers separate. Then separate the layers as discussed in the first part of the Appendix.

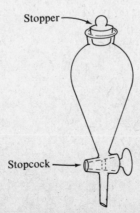

Stopper

Stopcock

Fig. 21.1 *Separatory funnel.*

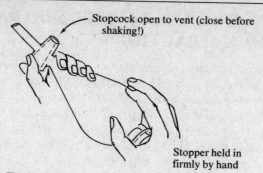

Stopcock open to vent (close before shaking!)

Stopper held in firmly by hand

Fig. 21.2 *Venting separatory funnel during extraction.*

EXPERIMENT

22

Determination of the Boiling Point of a Mixture of Liquids

I. INTRODUCTION

A. General

This experiment is designed to determine the boiling point of a mixture of liquids. The object is to study a property of solutions that is a measure of the strength and nature of the interaction between the different molecular species present. In order to understand the results of this experiment one must first realize that all condensed states of matter have a finite *vapor pressure* which almost always increases with increasing temperature, as shown in Figure 22.1. When the vapor pressure of a liquid is equal to the pressure of the surrounding atmosphere (point B on Fig. 22.1), bubbles may form in the solution interior and the solution is seen to boil. Assuming that the composition of the solution is held constant during the

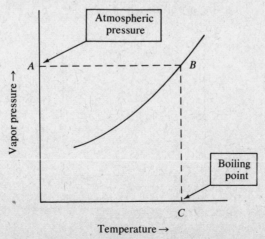

Fig. 22.1 *Vapor pressure/temperature plot.*

boiling process, the addition of more heat to the solution will not result in an increase of temperature but will cause more vigorous boiling. The main point of this experiment is that the boiling point (point C of Fig. 22.1) determines the temperature at which the vapor pressure of the solution equals atmospheric pressure.

The vapor pressure of a mixture of two liquids (which we will call A and B) usually lies somewhere between the vapor pressure of pure A (denoted p_A^o) and the vapor pressure of pure B (p_B^o). For some solutions the vapor pressure of the mixture follows the *ideal solution law (Raoult's Law)*, which is

$$P_{\text{mixture}} = p_A^o X_A + p_B^o X_B \tag{1}$$

where X_A = mole fraction of A

$\quad\quad\; X_B$ = mole fraction of B.

Basically, Raoult's Law says that the total pressure of a mixture of liquids is dependent upon the proportion of the molecules of each kind present in the mixture. For an ideal solution the total vapor pressure should be the sum of the contributions of each kind of molecule, which is proportional to the fraction of that kind of molecule in the liquid. This relationship for an ideal solution is given in Figure 22.2.

Often there are deviations from the ideal solution law. We may differentiate three cases that describe solution behavior based upon the relative magnitude of the forces acting between the molecules present in the mixture:

1. The attractive forces of A for A, and B for B are essentially equal to A for B. If we denote these forces by $F_{(A-A)}$, $F_{(B-B)}$, and $F_{(A-B)}$, then $F_{(A-A)} \approx F_{(B-B)} \approx F_{(A-B)}$.
  In this case Raoult's Law is valid.

2. The A-A attraction is greater than the A-B or B-B attraction, that is, $F_{(A-A)} > F_{(A-B)} \approx F_{(B-B)}$.

In this case vapor pressure of the mixture will be greater than that predicted by Raoult's Law. The reason is that the B molecules are "squeezed" out of solution by the A molecules clustering around each other.

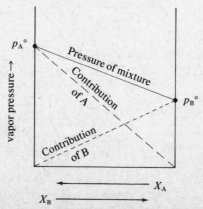

Fig. 22.2   *Vapor pressure composition diagram of the ideal system $X_B - X_A$.*

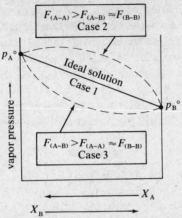

Fig. 22.3   *Vapor pressure composition diagram of the non-ideal system $X_B - X_A$.*

3. The A-B attraction is stronger than the A-A or B-B attraction, i.e., $F_{(A-B)} > F_{(A-A)} \approx F_{(B-B)}$.

In this case the vapor pressure is lower than that predicted by Raoult's Law, as the A-B attractive forces are stronger than those of the pure A or B liquid.

These possibilities are illustrated in Figure 22.3.

In this experiment we do not measure vapor pressure of the solution directly, but rather its boiling point (T_b). However, the discussion of the variation of the solution vapor pressure of a mixture with composition applies equally well to the variation of T_b with solution composition. Consider the same 3 above:

1. If the solution is ideal and $F_{(A-A)} \approx F_{(A-B)} \approx F_{(B-B)}$, then a plot of T_b of a mixture as a function of mole fraction lies essentially on a straight line joining the boiling point of the pure A(T_A) or pure B(T_B) (see Fig. 22.4).

2. If $F_{(A-A)} > F_{(A-B)} \approx F_{(B-B)}$, then the T_b versus mole fraction plot will be below

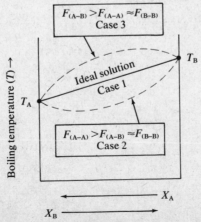

Fig. 22.4   *Boiling point composition diagram of the system $X_B - X_A$.*

the straight line joining T_A and T_B (see Fig. 22.4). In this case the vapor pressure of this mixture is higher than an ideal mixture (see Fig. 22.3), so the temperature at which the solution vapor pressure equals atmospheric pressure is lower than one would expect from the ideal solution law.

3. If $F_{(A-B)} > F_{(A-A)} \approx F_{(B-B)}$, then the T_b versus mole fraction plot will be above the straight line joining T_A and T_B (see Fig. 22.4). This result is found because in this case the vapor pressure of the solution is lower than that of the ideal solution, so a higher temperature for boiling is required.

B. Discussion of Hydrogen Bonding in Liquids

The molecules whose solutions are to be studied are listed in Table 22.1.

Water and the three alcohols form very strong *hydrogen bonds* with themselves (structures I and II) or each other (structure III):

$$R = -CH_3, -C_2H_5, -CH(CH_3)_2$$

I   II   III

The strength of the hydrogen bond is larger for the water-water interaction than for the interaction of alcohol molecules. The strength of the alcoholic hydrogen bond is approximately independent of the type of R groups. Acetone can form hydrogen bonds only if it is an acceptor; it is capable of forming a hydrogen bond with sufficiently acidic hydrogen atoms such as is shown in structures IV, V, and VI.

Table 22.1
Properties of Solvents To Be Studied

Name	Density 25°C g/ml	Structure	$T_b (P = 1\ atm)$ (°C)
water	1.00	H O H	100.0
methyl alcohol (methanol)	0.79	H_3C O H	64.7
ethyl alcohol (ethanol)	0.79	H_3C-H_2C O H	78.5
isopropyl alcohol	0.80	$(CH_3)_2HC$ O H	82.3
acetone	0.79	H_3C $\overset{O}{\underset{\parallel}{C}}$ CH_3	56.3
chloroform	1.18	Cl_3C-H	61.3

$$H_3C\diagdown$$
$$\qquad C=O-----H\diagdown$$
$$H_3C\diagup \qquad\qquad O$$
$$\qquad\qquad\qquad H$$

IV

$$H_3C\diagdown$$
$$\qquad C=O-----H-O$$
$$H_3C\diagup \qquad\qquad\qquad |$$
$$\qquad\qquad\qquad\qquad R$$

V

$$H_3C\diagdown$$
$$\qquad\qquad C=O-----H-C(Cl)_3$$
$$H_3C\diagup$$

VI

However, the acetone does not form hydrogen bonds with itself because the methyl hydrogen atoms ($-CH_3$) are not sufficiently acidic. The chloroform molecule ($CHCl_3$) can form hydrogen bonds only if it is a donor; it is capable of forming hydrogen bonds with any part of a molecule with a high electron density such as the oxygen atom in acetone (VI), or in an alcohol (VII); see VI and VII.

$$\qquad\qquad\qquad\qquad\qquad H$$
$$\qquad\qquad\qquad\qquad\qquad\diagup$$
$$(Cl)_3C-H-----O$$
$$\qquad\qquad\qquad\qquad\qquad\diagdown$$
$$\qquad\qquad\qquad\qquad\qquad R$$

VII

Various combinations of the above molecules are adequate to cover the three main solution cases discussed in section IA of this experimental description.

In this experiment, mixtures of two pairs of these molecules will be studied with the boiling temperatures being determined for various weight fractions. The weight fraction will be determined by measuring out known volumes of the pure liquids and converting these volumes to corresponding weights of the substances using the densities listed in Table 22.1. Convert the weight fraction into the corresponding mole fraction (X) using the relationships shown in Equations 2 and 3.

$$N_A \text{ (number of moles of A)} = \frac{\text{weight of A}}{\text{molecular weight of A}} \qquad (2)$$

$$X_A \text{ (mole fraction of A)} = \frac{N_A}{N_A + N_B} \qquad (3)$$

Two points of caution:

1. The liquids to be studied often contain a significant percentage of water; consequently the boiling point of the "pure" solvent will not necessarily be the same as those in Table 22.1. Also the atmospheric pressure may not be exactly one atmosphere or your thermometer may not be accurate. If the boiling point of the pure solvent deviates by more than 3°C from the value listed in Table 22.1 it should be rejected and a fresh sample used.

2. All the solvents studied (except for water) are highly flammable. Follow all procedures carefully and keep all chemicals well separated from the Bunsen burner flame.

II. PROCEDURE

The laboratory instructor will assign each student a pair of solvents to investigate.

The basic apparatus to be used is shown in Figure 22.5. The various parts of the apparatus each have a definite purpose:

a. The water bath provides for even, rapid heating of the mixture. In addition, it avoids heating of volatile and inflammable substances over an open flame. If the test tube were to crack, the highly flammable solvents will be harmlessly diluted in the water bath. Keep drafts away from the burner.

b. The glass tube condenser is very important. The vapors that are evolved from the boiling mixture condense on the walls of the glass tube and are returned to the liquid. This prevents (1) changes in the composition of the liquid during boiling, (2) fires arising from ignition of the flammable vapors, and (3) asphyxiation of the experimenter. Obviously, it is also important that the cork fits snugly into the test tube.

c. The thermometer is suspended inside the glass condenser by a wire or string. The bulb of the thermometer must extend at least one inch into the solution.

The basic experimental procedure to determine the boiling point of each mixture prepared is as follows:

a. The water bath should be heated to approximately 10°C above the expected boiling point of the mixture. Use a 400 ml beaker nearly filled.

b. Each individual mixture is added to the test tube *along with 2–3 boiling chips* and the test tube placed in the water bath. The level of the organic mixture should be *below the water level* in the bath.

> ☛ **CAUTION:** The boiling chips are essential to ensure proper boiling of the solution. Without them the solution may superheat and "bump."

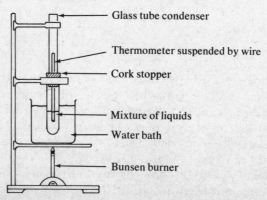

Fig. 22.5 *A schematic diagram of the apparatus used to establish the boiling point of a liquid.*

c. The glass condenser is quickly put in place.

d. The temperature should be monitored until it becomes constant. A slight temperature increase will be observed after the solution begins boiling, as the bulk of the solution comes to temperature. Also vapor will be observed condensing in the glass condenser. DO NOT ALLOW THIS VAPOR TO RISE HIGHER THAN ABOUT 12 in. ABOVE THE CORK. A gentle boil is best for the experiment. Record the equilibrium temperature of each solution in your notebook.

e. Dispose of each mixture carefully according to the instructions of the laboratory supervisor.

At least 8 data points should be taken for the solution pair studied, as well as the boiling points of the pure solvents. In case pure water is one component of the mixture assigned to you, the boiling point will have to be determined by direct heating. The mixtures are to be prepared in the 50 ml graduated cylinder. The desired volume of each solvent is added to make total volume 25–30 ml (add the larger volume first). Record in your notebook the volume of each solvent for a particular mixture and the boiling point as indicated in Table 22.2. X_A should take the *approximate* values 0, 0.10, 0.3, 0.5, 0.7, 0.9, 1.0. In the presentation of the results (see section III), the boiling point of the mixture will be plotted versus X_A. TAKE THE IDENTITY OF MOLECULE A TO BE THE SUBSTANCE WITH THE HIGHEST BOILING POINT, which will make your results appear qualitatively like those in Figure 22.4.

Table 22.2

Volume A	Weight A[1]	Volume B	Weight B[1]	N_A^2	X_A	T_b

[1]Calculated from the density of the substance.
[2]Calculated from the weight and the molecular weight of the substance (see Eq. 2).

III. DATA ANALYSIS

The report for this experiment consists of a plot of T_b versus X_A. The graph should be exactly like the graph on the next page. Label all curves carefully.

While it is not required for the report, you can deduce which of the three cases described in section IA best describes the solution pair(s) you have studied. You should be able to conclude from your results whether or not the mixture you investigated is ideal and, if not, what the nature of the hydrogen bonds between A and B is. Use the discussion section IIB to guide your conclusions.

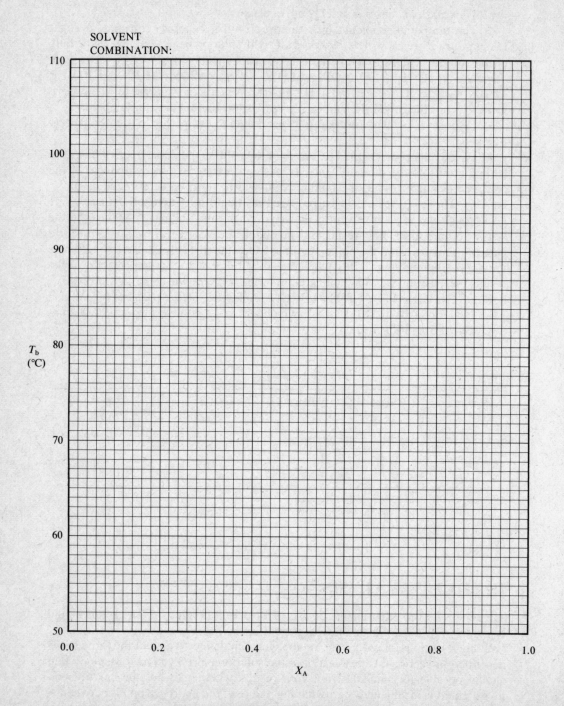

SOLVENT
COMBINATION:

T_b
(°C)

X_A

Fig. 22.6 *Graph for data of Experiment 22.*

IV. ERROR ANALYSIS

1. Given the general shape of your boiling point versus mole fraction curve, what kind of error would result from allowing the vapor to rise too high in the condenser?

2. Estimate the error in the calculated mole fraction that will result from ± 1 ml in the volume of each liquid (the error will be dependent on density) for the following volume combination: $V_A = 5$ ml, $V_B = 25$ ml; $V_A = 15$ ml, $V_B = 15$ ml; $V_A = 25$ ml, $V_B = 5$ ml.

3. What will be the effect on the boiling point–composition curve of a small leak at the cork-test tube joint? Will T_b tend to be high or low?

4. Suppose the densities of all liquids decrease by 10 percent between the room temperature and the boiling point. What effect does this have on the calculated mole fractions?

NAME   DATE

SECTION

SELF-STUDY QUESTIONS

1. What is the relation between the boiling point of a liquid and its vapor pressure?

2. What is Raoult's Law? Sketch the boiling point versus mole fraction for a pair of liquids obeying Raoult's Law.

3. What is the molecular interpretation of a positive deviation of T_b versus X_A from Raoult's Law? What is the interpretation of a negative deviation?

4. What is the purpose of the boiling chips in this experiment?

5. What is the purpose of the glass tube condenser?

6. Why is the mixture heated by a water bath rather than directly by the flame?

7. Why should you wait several minutes after boiling commences before taking a T_b measurement?

8. Name several things that can happen if the cork is not tight in the test tube or the vapor rises too far in the condenser or actually escapes the condenser.

EXPERIMENT

23

The Vapor Pressure of Liquids

I. INTRODUCTION

Regardless of their chemical composition all liquids show some tendency to evaporate, that is, to change from the liquid to the vapor state. Moreover, liquids evaporate at different rates which are dependent upon the temperature of the liquid, the number of molecules of the liquid in the immediate vicinity of the liquid surface, and the surface area of the liquid.

Liquids do not appear to evaporate in closed containers. This does not imply that molecules do not leave the surface of the liquid. The average kinetic energy of the molecules in the sample depends only upon the temperature and not upon whether the container is closed.

If water—or any liquid—is placed in a closed container as shown in Figure 23.1 some water molecules escape from the liquid and exist admixed with the air above the liquid. At any given temperature the quantity of water that can exist in the gas above the liquid is limited. The gaseous molecules strike other molecules and the walls of the vessel in a random manner, eventually strike the surface of that liquid, and again come under the influence of the molecules that make up the liquid. Fi-

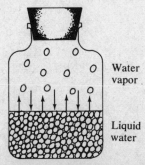

Water vapor

Liquid water

Fig. 23.1 *Liquid-vapor equilibrium in a closed container.*

nally, a situation develops where the number of molecules re-entering the liquid per unit time is the same as the number leaving. Thus, there are two physical processes.

$$liquid \longrightarrow gas \tag{1}$$

$$gas \longrightarrow liquid \tag{2}$$

occurring in the vessel simultaneously and at the same rate; these counterbalance each other so that there is no net change even though both processes continue to occur. This condition is known as a *dynamic equilibrium* and is not limited to physical processes. An increase in the temperature of liquid water causes more water molecules to leave the surface, increasing the concentration of gaseous molecules in the vessel, and a new state of equilibrium is established. The pressure exerted by a gas in equilibrium with the corresponding liquid at a fixed temperature is called the *vapor pressure* of the liquid at that temperature. All liquids exhibit vapor pressure curves which have the general appearance of that shown in Figure 23.2, with more or less steeply rising slopes. Thus, in Figure 23.2, liquid A has a higher vapor pressure (at all temperatures) than liquid B. It should be apparent that the vapor pressure curve is a characteristic property of a liquid.

A mathematical expression that describes vapor pressure data for most liquids is given in Equation 3

$$\log p = \frac{-\Delta H}{2.303R} \frac{1}{T} + B \tag{3}$$

where p is the vapor pressure at temperature T (expressed in the absolute scale)

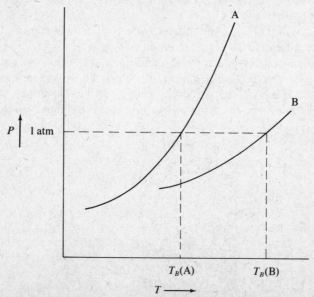

Fig. 23.2 *Vapor pressure versus temperature curves for typical liquids.*

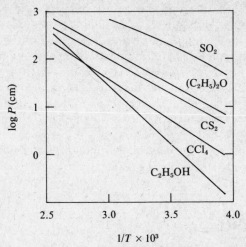

Fig. 23.3 *Vapor pressure curves plotted according to the Clausius–Claperyon expression.*

and R is the gas constant usually expressed in calories (R = 1.987 cal/mole °K). Note that Equation 3 is of the form of a straight line

$$y = mx + b \tag{4}$$

where y is identified with the term $\log p$ and x with $1/T$; historically, Equation 3 is known as the *Clausius-Claperyon equation,* and the constant, ΔH, is identified with the enthalpy of vaporization of the liquid (the enthalpy of evaporation is sometimes called the *heat of vaporization*). The constant B, although it has physical meaning, is not important for our purposes here. Figure 23.3 shows a number of vapor pressure curves plotted according to the Clausius-Claperyon expression. The enthalpy of vaporization reflects the differences in the forces of interaction between the molecules in the liquid state. Note that the units of pressure used in the Clausius-Claperyon equation are unimportant because all the information comes from the slope of the plot.

The escaping tendency of molecules at the surface of a liquid is opposed by the pressure of the gas above the liquid. The temperature at which the liquid exerts a vapor pressure slightly greater than atmospheric pressure is called the *boiling point* of the liquid. This is the temperature at which very many molecules throughout the liquid have achieved a kinetic energy large enough to escape from the attractive forces of the other molecules that make up the liquid. These energetic molecules form bubbles containing the substance in the gaseous state which escape against the pressure of the atmosphere. It should be apparent that since "atmospheric pressure" varies in the real world, boiling points will vary. Thus, the normal boiling point of a liquid is defined as the temperature at which its vapor pressure is equal to 1 atmosphere (760 torr); recall that 1 atmosphere is the usual standard pressure.

The boiling point of a liquid can be extracted from vapor pressure data quite simply. If the data are plotted in the usual way (Fig. 23.2), the point at which the

"1 atm line" intersects the vapor pressure curve is the boiling point of the substance. For example, in Figure 23.2 the boiling points, T_B, of the substances A and B are indicated on the temperature axis. It should now be obvious why vapor pressure and boiling points as well as heats of vaporization are characteristic qualities of liquids.

If the vapor pressure data are given in terms of the Clausius-Claperyon equation (Eq. 3), the normal boiling point can be obtained by substituting one atmosphere (expressed in the appropriate units) for p in the "log p" term and solving the expression for T.

In this experiment you will determine vapor pressure data for an unknown liquid and determine its normal boiling point.

II. PROCEDURE

1. Set up the apparatus as shown in Figure 23.4. Make sure that the barrel of the syringe is *lightly* greased and moves easily. Check that the test tube and stopper and the Luer-lok fit tightly enough to keep water out when submerged. Be sure you have a stirring rod for the water bath (*do not* use the thermometer for a stirring rod). Record the barometric pressure (the instructor will either give you this information or show you the location of a barometer). The volume of the test tube (with the stopper in place) can be determined by filling the test tube with water and

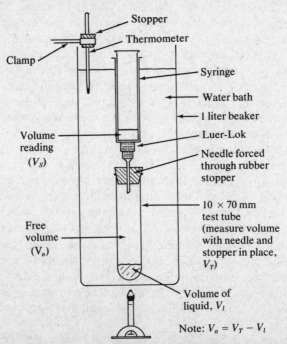

Fig. 23.4 *Apparatus for Experiment 23.*

then pouring the water into an appropriate graduated cylinder (10 ml would be best). The volume can also be determined by weighing the water and looking up the density of water at room temperature (see Experiment 1). The free volume of the test tube (denoted V_o) will be this volume minus 5 ml, the volume of liquid added.

2. Add a 5 ml sample of the unknown liquid to the test tube. Disconnect the syringe from the needle and pull the barrel to the 5 ml marker on the syringe. Fit the stopper snugly into the test tube, refit the syringe into the needle Luer-lok, and clamp the whole arrangement in place with all the test tube and most of the syringe emersed (see Fig. 23.4). Stir the water bath several times until thermal equilibrium is achieved (5–10 minutes). Record the temperature of the bath and the volume reading of the syringe (the syringe reading will be denoted as V_s). The value of $V_o + V_s$ will be defined as V_c in the Data Analysis section.

3. Heat the bath, with stirring, to approximately 5°–7° above the initial temperature. After 5–10 minutes record the temperature of the water bath and the new volume on the syringe. Warm the heat bath another 5°–7° and repeat until you have 5–7 data points. The following are convenient headings for the columns in your data table (quantities marked with asterisk (*) are the recorded experimental data; the other quantities are calculated in the data analysis).

$$|T_i(°C)^*|T_i(°K)|1/T_i(°K^{-1})|(V_s)_i(ml)^*|V_i = V_o + (V_s)_i(liter)|P(X)_i(atm)|log\ P(X)_i|$$

III. DATA ANALYSIS

We must remember 3 important points when analyzing the data. (1) We would like to obtain the vapor *pressure* of the liquid but the primary data which we collect is the *volume* of the system. (2) As we heat the system, the volume increase arises from (a) the thermal expansion of the air originally in the apparatus and (b) the increase in the number of unknown molecules which have evaporated from the liquid. (3) The experiment is performed at constant pressure, viz., the atmospheric pressure prevailing in the laboratory room when you did the experiment.

Let us begin the analysis by defining a few symbols.

V_c is the volume of air trapped in the apparatus when you first assembled it. This volume represents a certain number of air molecules which are always present during all the measurements. For V_c, use the initial syringe volume reading (V_s) plus the free volume in the test tube (V_o) (see Fig. 23.4), i.e., $V_c = V_s + V_o$.

$(V_s)_i$ is the volume reading of the syringe at a temperature T_i. $V_i = (V_s)_i + V_o$.

P_R is the prevailing atmospheric pressure in the laboratory at the time of the experiment.

T_1 is room temperature, i.e., the first temperature reading you made.

T_i is the temperature of the water bath; T_i will be some temperature higher than the initial room temperature.

1. The air originally trapped in the apparatus must obey the ideal gas law

$$V_c P_R = nRT_1 \tag{5}$$

Rearranging Equation 5 for the number of moles of air present we obtain

$$n = \frac{V_c P_R}{RT_1} \tag{6}$$

NOTE: This analysis assumes that the vapor pressure of the liquid may be neglected with respect to atmospheric pressure. For sufficiently volatile liquids an ice bath may be required for T_1.

---

2. At any temperature T_i, the pressure of the system arises from the original air as it has expanded and the evaporated liquid. We assume the validity of Dalton's Law of partial pressures.

$$P_R = P(air)_i + P(X)_i \tag{7}$$

where P_R is the prevailing atmospheric pressure (which is exerted on one side of the syringe barrel), $P(air)_i$ is the pressure of the original air in the system as it has expanded due to heating, and $P(X)_i$ is the pressure of the unknown in equilibrium with its liquid state. $P(X)_i$ is the vapor pressure of the liquid at temperature T_i.

3. We assume that the air present in the system at the higher temperatures still obeys the ideal gas law (Eq. 8).

$$P(air)_i V_i = nRT_i \tag{8}$$

The number of moles of air has not changed and is still given by Equation 6 which can be substituted into Equation 8 to give

$$V_i P(air)_i = \left(\frac{V_c P_R}{RT_1}\right) RT_i \tag{9}$$

Solving Equation 9 for $P(air)_i$ and canceling factors where appropriate we obtain

$$P(air)_i = \frac{V_c T_i P_R}{V_i T_1}. \tag{10}$$

4. Equation 10 can be substituted into Equation 8 to give

$$P_R = \frac{V_c T_i P_R}{V_i T_1} + P(X)_i. \tag{11}$$

Equation 11 can be solved for the desired quantity $P(X)_i$

$$P(X)_i = P_R - \frac{V_c T_i P_R}{V_i T_1} \tag{12}$$

or

$$P(X)_i = P_R \left[1 - \frac{V_c T_i}{V_i T_1}\right]. \tag{13}$$

All the quantities on the right side of Equation 13 are experimentally known and $P(X)_i$ can be calculated at each temperature. Refer to your data table for the experimental and calculated quantities for each trial.

Prepare a plot of log $P(X)_i$ as a function of $1/T$ for your data (Fig. 23.5); estimate the normal boiling point of the liquid. Fit your data to the Clausius-Claperyon equation (Eq. 3) (use a least-squares method, see Appendix 2) and obtain the heat of vaporization of the unknown.

Report the numerical value of the normal boiling point of the liquid (in °C), and its heat of vaporization (in kcal/mole).

IV. ERROR ANALYSIS

For the following, estimate the uncertainty in the calculated vapor pressure of your unknown from each error.

1. The actual atmospheric pressure is 10 percent higher than you recorded (barometer broken).

2. The syringe is "sticky" such that the internal pressure is actually 10 percent higher than atmospheric pressure when the syringe stops moving.

3. The temperature you record at 35° is off by ±1°C (assume T_1 = 23°C).

4. The value of V_i is off by ±1 ml for T_i = 35° (pick a typical data point in your table and your values of V_o, T_1, P_R).

5. The value of V_o is off by ±1 ml (use your data for some typical T_i).

6. Considering all of the above and the likely magnitude of your experimental error, estimate the ''error bars'' on the values of $\log P(X)_i$ in your plot. Estimate the resulting uncertainty in ΔH_{vap} derived from your data.

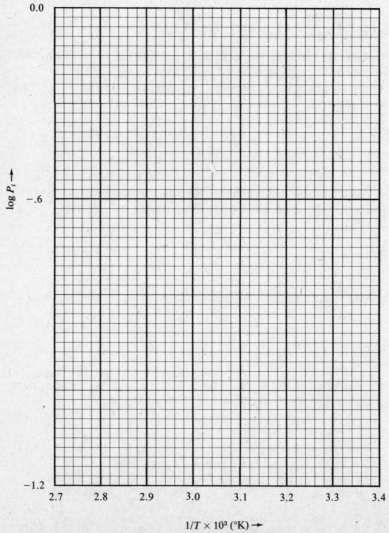

Fig. 23.5 *Graph for data of Experiment 23.*

SELF-STUDY QUESTIONS

1. What is meant by the term *vapor pressure?*

2. If the enthalpy of vaporization (ΔH_{vap}) is large (it is always positive) for a given liquid will the vapor pressure be large or small? What does a large ΔH_{vap} imply about the strength of the interaction between the molecules in the liquid phase?

3. If the ΔH_{vap} is known for a liquid, how can one predict (with reasonable accuracy) the boiling point for that liquid if the vapor pressure at one temperature is known?

4. Sketch the experimental apparatus used to measure the changes in volume occupied by the air plus liquid vapor.

5. If the vapor pressure of the liquid were essentially zero at all temperatures, what would be the equation for the measured volume (V_i) at different temperatures in terms of V_c (the original volume) and T_1 (the original temperature).

EXPERIMENT

24

Fractional Distillation: Separation of Two Volatile Components of a Mixture

I. INTRODUCTION

One of the most useful techniques for purifying mixtures of volatile components is fractional distillation, in which the difference in the vapor pressure of the various components is exploited to effect a separation. It is the objective of this experiment to fractionate a mixture of two volatile liquids into pure components. The experimental set-up for fractional distillation is illustrated in Figure 24.1. As can be seen in that figure, a round-bottomed flask is used to hold the original "charge" of mixed solvents. The mixture is boiled and the vapor is cooled in the water jacketed condenser and then collected. A thermometer indicates the temperature of the vapor phase that is being condensed. The most important feature of the arrangement shown in Figure 24.1 is the *distillation head,* which is filled with glass beads, glass helices, or some other equivalent material that provides a large surface area for the condensation of vapor and re-evaporation of the liquid. The distillation head also allows a temperature gradient to be established along its length. To understand why this feature of the apparatus is important we must consider the relationship between the composition of the vapor phase for a given composition of the liquid phase. For the case of an ideal solution of volatile components A and B the vapor pressure is given by Raoult's Law, i.e.,

$$\begin{aligned} P_{\text{total}} &= p_A + p_B \\ &= p_A^\circ X_A^L + p_B^\circ X_B^L \\ &= p_A^\circ X_A^L + p_B^\circ (1 - X_A^L) \end{aligned} \tag{1}$$

where p_A°, p_B° are the vapor pressures of pure A and B (at the given temperature) and X_A^L, X_B^L are the mole fractions of components A and B in the liquid (note that $X_A^L + X_B^L = 1$, so $X_B^L = 1 - X_A^L$). The mole fraction of component A in the vapor phase is given by

$$X_A^V = \frac{p_A}{p_{\text{total}}} = \frac{X_A^L p_A^\circ}{p_B^\circ + (p_A^\circ - p_B^\circ) X_A^L} \tag{2}$$

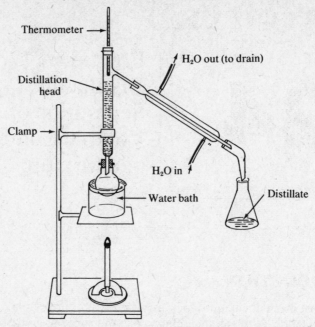

Fig. 24.1 *Typical experimental set-up for fractional distillation.*

and $X_B^V = 1 - X_A^V$. While it is not particularly obvious from casual inspection of Equation 2, the composition of the vapor phase is richer in the more volatile component. This information can be represented by a plot like that shown in Figure 24.2 in which vapor pressure is plotted as a function of X_A^L at constant temperature. One must be careful in interpreting this plot. For example, if a solution has a composition $X_A^L = 0.3$, then the total vapor pressure will be given by the intercept of the line *a-b* on the L (= liquid) line (at $X_A^V = 0.7$ in Fig. 24.2). The intercept of the line *c-b-d* on the V (= vapor) line yields the composition of the vapor, X_A^V, in equilibrium with the liquid phase with $X_A^L = 0.3$. As can be seen $X_A^V > X_A^L$, so the vapor is enriched in A, the more volatile component (from Fig. 24.2 it is seen that $p_A^\circ > p_B^\circ$). For our present purposes of distillation the total vapor pressure will be maintained at a constant value of atmospheric pressure by heating to the boiling point, T_b. In Figure 24.3 is a plot of T_b versus X_A which is similar to Figure 24.2 except that total pressure is constant and the temperature is variable. In this case the vapor at a given T_b is richer in the more volatile component (as before, assumed to be component A). Suppose that it is possible to take the vapor at point *q* which is in equilibrium with boiling liquid at temperature $(T_b)_q$ and liquid composition $(X_A)_I^L$ and cool it to T_r. At that temperature liquid of composition $(X_A)_I$ condenses and maintains an equilibrium with vapor of composition $(X_A^V)_{II}$. It is this series of steps that the distillation head accomplishes. The vapor heats the length of the packing material (glass beads, etc.) according to the boiling point of the mixture in equilibrium at that point. In the ideal case the bottom part of the distillation head is at the boiling point of the mixture and the top

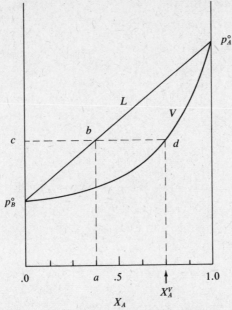

Fig. 24.2 *Typical vapor pressure vs. composition plot for an ideal solution.*

portion is at the boiling point of the component with the lowest boiling point. In the ideal fractional distillation the temperature registered at the top of the distillation head will correspond to one pure component until it has been removed from the mixture, then the temperature will rise rapidly to the boiling point of the second component. An actual distillation will rarely work quite that well, but an excellent separation of solution components is generally possible with care. A special problem that sometimes arises is the formation of an *azeotropic mixture,* which is a mixture with a constant composition of liquid and vapor, such that no further separation of the two components can occur. This is discussed briefly in Appendix 24.

II. PROCEDURE

As was stated in the first section, for this experiment you will be issued an unknown mixture of two volatile liquids, and by means of fractional distillation separate them into the two pure components and report the *volume fraction* of each component (see section 4, p. 239). The procedure to be followed for this experiment is extremely simple, but there are two important points that should be noted:

1. All connections in the apparatus in Figure 24.1 should be snug so that the vapor of the distillate does not escape into the room. In general the chemicals used are *noxious and flammable*.

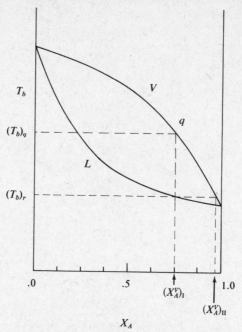

Fig. 24.3 *Typical boiling point vs. composition plot.*

2. In order for this distillation to be effective (and hence your results to be accurate) the rate at which the distillate is collected must not be too high (a few seconds between drops is about right). Hence the distillation should be started promptly at the beginning of the laboratory period and it should be expected that the distillation will take most of one laboratory period.

Detailed Procedure

1. Prepare the distillation apparatus as in Figure 24.1. Check all joints. Add a few boiling chips. Place 50 ml of the unknown solution in the round-bottomed flask.

NOTE: You must be accurate to within ±1 ml in the amount of unknown placed in the flask.

Besides measuring the volume carefully in your graduated cylinder (which must be dry), let the graduated cylinder drain into the flask thoroughly.

2. Heat the water bath fairly strongly until boiling commences in the round-bottomed flask. Once vapor can be seen condensing about halfway up the distillation head the rate of heating should be carefully controlled such that the vapor rises slowly up the distillation head. The thermometer will register a sharp rise of temperature when vapor reaches the top of the head and condensate will begin to be collected. Note the temperature at which the vapor first begins to be collected, and take readings every minute until the distillation temperature has equilibrated.

3. As was mentioned in the Introduction, the temperature should remain about constant until all of the more volatile component has been removed, at which point the temperature of the vapor will rise fairly quickly to the boiling point of the second component. Hence the temperature must be monitored continuously to judge when a sudden temperature increase occurs. (Don't wander off and leave the apparatus unattended!) When a definite temperature increase is observed, stop the distillation immediately by:

a. turning off the bunsen burner;

b. removing the water bath from contact with the round-bottomed flask (but see CAUTION).

☛ **CAUTION:** Step b can be dangerous unless appropriate tongs or insulated towels and/or gloves are available. Do *not* attempt this procedure unless it has been demonstrated by the laboratory instructor.

---

4. Allow all portions of the apparatus to cool to nearly room temperature. Measure the volume of distillate (= V_D), and the volume of liquid remaining in the round-bottomed flask (= V_R). If time allows, a second distillation can be carried out with the same unknown solution.

III. DATA ANALYSIS

Three measured values should be obtained: (1) V_D, volume of distillate obtained, (2) V_R, volume of distillate in flask, and (3) the temperature at which the distillate was recovered. The latter quantity should be reported, and the first two quantities should be reported as follows:

a. Fraction recovery of all liquids = $(V_D + V_R)/50$ (this assumes that the original amount of the unknown liquid used was 50 ml).

b. Volume fraction of distillate = $V_D/50$.

IV. ERROR ANALYSIS

For the following consider the qualitative effect of the error on $V_D/50$ or $(V_D + V_R)/50$, i.e., does the error stated make the quantity too large, too small, or have no effect?

1. An error in the initial volume of solution by +1 ml (i.e., putting 51 ml in the round-bottom flask when the experimentalist thinks he or she put in 50 ml).

2. Temperature reading errors of $\pm 1°C$.

3. The volume of liquid that adsorbs onto the surface of the head or condenser.

4. Evaporation losses from a loose connection between the round-bottomed flask and the distillation head.

5. Evaporation losses in the condenser because the water flow is inadequate to cool the distillate.

6. What will be the general effect of carrying out the distillation too fast? Will V_D or V_R be most affected?

7. What are some possible effects of leaving traces of water in the distillation apparatus?

NAME   **DATE**

SECTION

SELF-STUDY QUESTIONS

1. Sketch a fractional distillation apparatus (without referring to Fig. 24.1) and identify the various parts by name. What is the role of the distillation head?

2. The boiling point of liquid A is 62.5°C and the boiling point of liquid B is 84.0°C. Which component will be present in excess in the vapor phase which is in equilibrium with a solution of A and B where $X_A = X_B = 0.5$?

3. What is Raoult's Law?

4. Why does one add boiling chips to the mixture in the round-bottomed flask?

5. What is a good rate for the distillate to leave the condenser at? Why does it matter at what rate the distillation is done?

6. Why must the temperature at the top of the distillation head be monitored so carefully? Other than a bad grade, what is the likely effect of not paying attention to the temperature of the distillation?

APPENDIX 24. AZEOTROPIC MIXTURES

The discussion accompanying this experiment, and in particular Figures 24.2 and 24.3, was appropriate for ideal solutions (those solutions that obey Raoult's Law) or nearly ideal solutions (only slight deviations from Raoult's Law). Many mixtures are very non-ideal, usually because there exists some particular molecular interaction between the two components of the solution. (This is partially discussed in Experiment 22.) For some mixtures of volatile liquids the T_b versus composition curve may look like Figure 24.4(a) or 24.4(b). In both cases there is a point where the composition of the vapor and the liquid phases are identical, in which case no further purification by means of distillation is possible. It is these points in the composition curve that are referred to as *azeotropic mixtures*, which means "constant composition boiling mixtures." As far as the fractional distillation is concerned, the azeotropic mixture behaves like a pure component and one would not detect the presence of an azeotropic mixture unless the distillate or residue were analyzed separately.

There is one interesting difference between the behavior of the distillation in the types of properties displayed in Figure 24.4(a) and (b). In Figure 24.4(a) the azeotrope is lower boiling than either pure A or B, such that the first distillate is always the azeotropic mixture and the residue is pure A or pure B, depending on which side of the azeotropic point the original solution is on. In Figure 24.4(b) the azeotrope is higher boiling than either pure A or B, and the azeotropic mixture becomes the residue. Frequently the behavior of azeotropes is exploited by industrial or chemical processes.

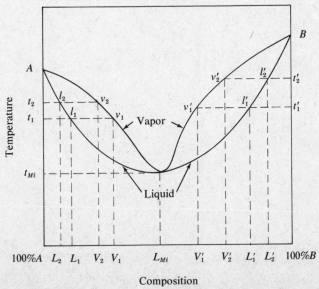

Fig. 24.4(a) T_b *vs. composition for low boiling azeotrope.*

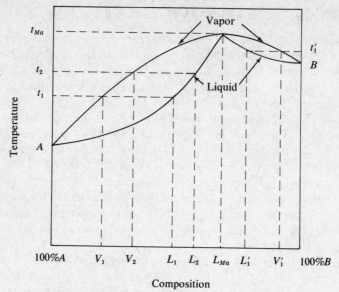

Fig. 24.4(b) *T_b vs. composition for high boiling azeotrope.*

EXPERIMENT

25

Colligative Properties and the Approximate Determination of Molecular Weight: Boiling Point Elevation

I. INTRODUCTION

Every pure substance has certain well-defined properties such as vapor pressure (at a given temperature) and melting point (at a given pressure). When a small amount of a substance (the solute) is dissolved in a larger amount of a second substance (the solvent), the properties of the solution (solvent and solute) are slightly different than the properties of the pure solvent. In many cases these changes in the solvent properties are independent of the molecular nature of the solvent or solute but rather depend only on the *ratio of the number of moles of solute to the number of moles of solvent*. Such properties are known as *colligative* (from the Latin, for "collective"). In this experiment the change in the boiling point of a solvent (ΔT_b) will be measured for a solution of a measured mass of solvent and solute, where the molecular weight of the solute will not be known. In the next experiment the change in the freezing point (ΔT_f) will be determined. A convenient relationship between the measured changes and the molecular weight of the unknown is

$$\Delta T_b = k_b \text{ (mass of solute/MW solute)(1000/mass of solvent)}$$
$$= k_b m_{\text{solute}} \tag{1}$$

$$\text{and} \quad \Delta T_f = k_f m_{\text{solute}} \tag{2}$$

where m_{solute} is known as the molality of the solution (defined as the number of moles of solute per 1000 g of solvent). In the present case it is the molecular weight of the solute that is unknown, and all other quantities in Equations 1 and 2 are measured; hence

Some typical values of k_b, k_f are listed in Table 25.1.

Equations 1 and 3a are to be used in this experiment without further analysis. A short discussion of the origin of these expressions will be given in Appendix 25. For the moment it is enough to state that Equations 1 and 3a will be valid when

1. there is no strong chemical or physical interaction between solvent and solute molecules (i.e., they form an "ideal solution");

2. the solute is not very volatile relative to the solvent (required for Eq. 1);

3. the solute is not very soluble in crystals of solvent, such that freezing the solution tends to produce crystals of pure solvent (required for Eq. 2).

In this experiment you will be issued a solute whose molecular weight is to be determined. The boiling point elevation of a solution ($\Delta T_b = T_b^{solution} - T_b^{solvent}$, T_b = boiling point) will be measured and Equation 1 will be used to determine the solute's molecular weight.

Table 25.1
Boiling Point Elevation and Freezing Point Depression Constants

Solvent	$T_b(°C)$	k_b	$T_f(°C)$	k_f
naphthalene	—	—	80.2	7.16
diphenylether	259	—	28	8.00
benzene	80.2	2.53	5.6	5.12
carbon tetrachloride	76.8	5.02	—	—
water	100	0.514	0.0	1.855
chloroform	61.2	3.63	—	—

$$\text{(MW solute)} = \frac{(k_b)(1000)}{\Delta T_b}\left(\frac{\text{mass of solute}}{\text{mass of solvent}}\right) \qquad (3a)$$

$$\text{(MW solute)} = \frac{(k_f)(1000)}{\Delta T_f}\left(\frac{\text{mass of solute}}{\text{mass of solvent}}\right). \qquad (3b)$$

II. PROCEDURE

The experimental arrangement will be like that of Experiment 22 and is illustrated in Figure 25.1. It is important for the stopper to fit snugly and for the heating to be controlled so that vapor only rises up 10–15 cm into the condenser tube. Minimizing the amount of liquid that vaporizes ensures that the composition of the liquid phase remains nearly identical to that of the original solution. *The use of boiling chips or pieces of glass to prevent superheating is essential.*

It is also essential that temperature readings for this experiment be accurate: even an error of ±0.2°C can be significant. After boiling has commenced and 1–3 minutes have passed for equilibration, several readings of the temperature at ~30 sec intervals should be taken.

The boiling point for the pure solvent will be measured first, and after a reliable value of T_b solvent is obtained, a small portion (0.4–0.6 g) of the solute will be added and T_b solution will be determined. A second portion of solute (0.2–0.4 g) will be added and the T_b solution determined. The apparent molecular weight of the solute from each experiment will be computed using Equation 3a.

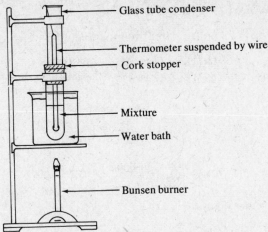

Glass tube condenser

Thermometer suspended by wire

Cork stopper

Mixture

Water bath

Bunsen burner

Fig. 25.1 *A schematic diagram of the apparatus used to establish the boiling point of a liquid.*

Detailed Procedure

1. Obtain a boiling point apparatus (like Fig. 25.1) and be sure all parts are clean and *dry* (rinse with ethanol or acetone and air-dry). Also clean and dry two small test tubes.

2. Obtain approximately 15 ml of solvent in a clean, dry 25 or 50 ml beaker. Your laboratory instructor will assign the particular solvent to be used. Also obtain approximately 1 g of the unknown solute in a small, dry beaker.

3. Weigh the beaker and solvent (to ± 0.01g) and pour the solvent into the large test tube on the boiling point apparatus. Weigh the empty beaker and calculate the mass of the solvent used. This mass will be designated as m_{solvent}.

4. Weigh the beaker and solute and then transfer a portion of the solute to one of the small, dry test tubes mentioned in step 1 above until approximately 0.4–0.6 g of solute has been added (weigh to ± 0.002 g). Using the same procedure, add approximately 0.2–0.4 g of the solute to the second small, dry test tube. These two masses will be designated m_1 and m_2, respectively.

NOTE: This procedure is an example of weighing by difference (see Appendix 5).

---

5. Set up the boiling point apparatus as shown in Figure 25.1, and heat the water bath until the solvent begins to boil gently (don't be fooled by air being excluded from the boiling chips). After boiling has commenced, begin taking temperature readings at ~30 sec intervals until equilibrium value has been obtained. Take care that the vapor does not rise higher than 10–15 cm up the condenser tube. Once a value of T_b^{solvent} has been obtained, remove the apparatus from the water bath and allow to cool until only slightly above room temperature.

6. Add mass m_1 of the solute, weighed out in step 4, to the solvent, swirl to dissolve the solute, and reassemble the boiling point apparatus. Measure the boiling point as in step 5. This boiling point will be designated as $(T_b^{\text{solution}})_1$. Cool the solution, add m_2, and obtain the boiling point of this solution $(T_b^{\text{solution}})_2$.

III. DATA ANALYSIS

Calculate the boiling point and apparent solute molecular weight as follows:

$$(\Delta T_b)_1 = (T_b^{solution})_1 - T_b^{solvent}$$
$$(\Delta T_b)_2 = (T_b^{solution})_2 - T_b^{solvent}$$

and

$$(\text{MW solute})_1 = \frac{(1000k_b)}{(\Delta T_b)_1}\left(\frac{m_1}{m_{solvent}}\right)$$

$$(\text{MW solute})_2 = \frac{(1000k_b)}{(\Delta T_b)_2}\left(\frac{m_1 + m_2}{m_{solvent}}\right).$$

Report *both* molecular weights.

IV. ERROR ANALYSIS

Estimate the error in the value of the reported molecular weight of the unknown for each of the following types of error:

1. an error in ΔT_b of $\pm 0.5°C$;

2. an error of ± 0.01 g in the value of $m_{solvent}$;

3. an error of ± 0.002 g in the value of m_1 and m_2 (the mass of the two samples of solute).

4. A certain amount of solvent is constantly refluxing up onto the condenser walls and hence the volume of solvent in the solution is slightly reduced from that originally added. Suppose 1 ml of solvent is on the condenser walls. What effect will this have on the value of molecular weight obtained?

5. Estimate the overall magnitude of likely error in the molecular weight value reported, considering all of the above factors plus any others you believe might be significant.

NAME
DATE

SECTION

SELF-STUDY QUESTIONS

1. Very briefly describe why the addition of a non-volatile solute increases the boiling point of a liquid.

2. What is meant by the term *colligative property*?

3. Why is a condenser present in the apparatus in Figure 25.1. Why must the level of condensing vapors be kept fairly low in the condenser?

4. Describe briefly how the solute is weighed.

APPENDIX 25. THERMODYNAMIC DERIVATION OF BOILING POINT ELEVATION AND FREEZING POINT DEPRESSION FORMULAS

It is assumed that we are dealing with a liquid solution composed of a solvent with mole fraction X_1 and a solute with a mole fraction X_2, and furthermore $X_1 \gg X_2$. If the solute dissolves only in the liquid phase and is non-volatile, then any solid in equilibrium with the solution is pure solvent, and the vapor phase in contact is also pure solvent. Under these conditions the free energy per mole of the solvent in each phase is given by

$$G_{gas} = G^\circ_{gas} + RT \ln p = H^\circ_{gas} - TS^\circ_{gas} + RT \ln p \tag{1}$$

p = partial pressure of solvent in gas phase

$\ln(p)$ = natural logarithm of p, $(2.303) \log_{10}(p)$

$$G_{liquid} = G^\circ_{liquid} + RT \ln X_1 = H^\circ_{liquid} - TS^\circ_{liquid} + RT \ln X_1. \tag{2}$$

Note that Equation 2 is valid only for so-called "ideal solutions of pure solids."

$$G_{solid} = G^\circ_{solid} = H^\circ_{solid} - TS^\circ_{solid} \tag{3}$$

(There is no dilution term for the solid.) At equilibrium between any two phases the free energy per mole of all species must be equal,* so for the case of gas-liquid equilibrium

$$G^\circ_{liquid} + RT \ln X_1 = G^\circ_{gas} + RT \ln p. \tag{4}$$

(This is equivalent to Raoult's Law.) At the boiling point, $p = 1$ atm (this is essentially the definition of the normal boiling point), so the temperature at which boiling occurs may be estimated by using Equation 1 and Equation 2.

$$H^\circ_{gas} - T_b S^\circ_{gas} = H^\circ_{liquid} - T_b S^\circ_{liquid} + RT_b \ln X_1 \tag{5}$$

If $X_1 = 1$, Equation 5 could be used to compute the normal boiling point. In the present case, $X_1 = 1 - X_2$, and it can be shown that $\ln(1 - X_2) = -X_2$ if $X_2 \ll 1$. If we write $T_b = T^\circ_b + \Delta T_b$ (T°_b = normal boiling point), then we obtain the equation

$$\underset{(a)}{(H^\circ_{gas} - H^\circ_{liquid})} = \underset{(b)}{T^\circ_b (S^\circ_{gas} - S^\circ_{liquid})} - \underset{(c)}{RT^\circ_b X_2} \tag{6}$$

$$+ \underset{(d)}{\Delta T_b (S^\circ_{gas} - S^\circ_{liquid})} - \underset{(e)}{R\Delta T_b X_2}$$

where term e is assumed to be very small and can be neglected (i.e., ΔT_b and X_2 are both small). From Equation 5 for the case $X_1 = 1$, we find

$$H^\circ_{gas} - H^\circ_{liquid} = T^\circ_b (S^\circ_{gas} - S^\circ_{liquid}). \tag{7}$$

Using Equation 7 in Equation 6, we see that terms a and b cancel, and term d is

---

*If this were not true, the total free energy of the system could be lowered by transferring material from one phase to the other.

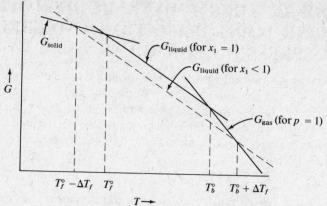

Fig. 25.2 *Representation of free energy per mole for different states of matter for pure solvent* $(X_1 = 1)$ *or solution* $(X_1 < 1)$.

$\Delta T_b (H^\circ_{gas} - H^\circ_{liquid})/T^\circ_b$,* so

$$0 = -RT^\circ_b X_2 + \Delta T_b (H^\circ_{gas} - H^\circ_{liquid})/T^\circ_b$$

or

$$\Delta T_b = \frac{R(T^\circ_b)^2 X_2}{H^\circ_{gas} - H^\circ_{liquid}} = \frac{R(T^\circ_b)^2}{\Delta H_{vap}} X_2. \qquad (8)$$

In order to obtain our final result in molalities, we always refer to 1000 g of solvent, or 1000 g/M_1 moles of solvent (M_1 = molecular weight of solvent) and if n_2 = number of moles of solute, then

$$X_2 = \frac{n_2}{n_2 + 1000 \text{ g}/M_1} \cong \frac{n_2}{1000} M_1 \qquad (9)$$

where it has been assumed that $n_2 \ll 1000/M_1$. The quantity $n_2/1000$ is known as the molality (m) of the solute, so

$$\Delta T_b = ((R(T^\circ_b)^2 M_1)/\Delta H_{vap}) m \qquad (10)$$

where the expression in parentheses is k_b. Note that only the properties of the solvent enter into this equation.

A similar argument can be constructed for the freezing point depression. At equilibrium

$$H^\circ_{solid} - T_f S^\circ_{solid} = H^\circ_{liquid} - T_f S^\circ_{liquid} + RT_f \ln X_1$$

and writing $T_f = T^\circ_f + \Delta T_f$, and $\ln X_1 = -X_2$ we obtain

$$\underset{(a)}{(H^\circ_{liquid} - H^\circ_{solid})} = \underset{(b)}{T^\circ_f(S^\circ_{liquid} - S^\circ_{solid})} + \underset{(c)}{RT^\circ_f X_2}$$

$$+ \underset{(d)}{\Delta T_f(S^\circ_{liquid} - S^\circ_{solid})} - \underset{(e)}{R \Delta T_f X_2} \qquad (11)$$

---

*This expression could just as well be left in terms of $S^\circ_{gas} - S^\circ_{liquid} = \Delta S^\circ_{vap}$ but most references will present the final results in terms of ΔH°_{vap}.

where e can be neglected for small ΔT_f°, X_2. As before, a and b cancel and, in d, $(S_{liquid}^\circ - S_{solid}^\circ) = \Delta H_{fusion}/T_b^\circ$ (ΔH_{fusion} is the enthalpy change in going from solid to liquid, $H_{liquid}^\circ - H_{solid}^\circ$), so

$$\Delta T_f = -((R(T_f^\circ)^2)/\Delta H_{fusion})X_2 = -((R(T_f^\circ)M_1)/\Delta H_{fusion})m.$$

Therefore, in this case, the shift in freezing is negative. In Equation 2 of the main text the absolute value of ΔT_f is used. The relations between the various free energy expressions in Equations 1–3 are sketched qualitatively in Figure 25.2.

EXPERIMENT

26

Colligative Properties and the Approximate Determination of Molecular Weight: Freezing Point Depression

I. INTRODUCTION

As was discussed in the Introduction section of Experiment 25, there are certain properties of solutions that are independent of the nature of the solute-solvent pair, assuming that the resulting solution is sufficiently close to ideality.* The property that is to be exploited in this experiment is the molal freezing point depression, which can be expressed by the equation

$$\Delta T_f = k_f \frac{(\text{moles of solute})}{(\text{mass of solvent})/1000}$$

$$= k_f \left(\frac{\text{mass of solute}}{\text{MW solute}} \right) \left(\frac{1000}{(\text{mass of solvent})} \right). \tag{1}$$

Solving for the molecular weight of the solute,

$$(\text{MW solute}) = \frac{k_f}{\Delta T_f}(1000)\frac{(\text{mass of solute})}{(\text{mass of solvent})}. \tag{2}$$

All quantities on the right side of Equation 2 are measured or known (see Table 25.1); hence, the molecular weight of the solute can be determined. In this experiment ΔT_f will be determined for two different masses of unknown solute added to a given mass of solvent, and an average molecular weight will be obtained.

Determination of a freezing point is somewhat more difficult than determination

---

*It is advised that the student read the Introduction section of Experiment 25 because Experiment 25 and Experiment 26 are closely related. It is also important to read Appendix 25 because of its relevance to this experiment.

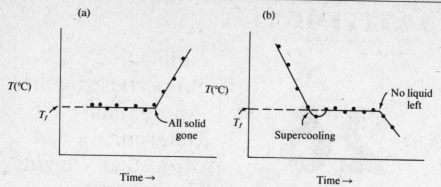

Fig. 26.1 *Heating and cooling curves for pure substances.*

of the boiling point, but is based on the same principle as the latter, namely, that the temperature of a system undergoing a phase change tends to remain constant, even if heat is absorbed or removed, until the phase change is complete. Consider, for example, a solid being heated to near its melting point. As heat is added the temperature rises until melting begins. At this temperature the addition of more heat results in more melting, but no temperature increase, until all solid is melted (this assumes that the mixture of solid is stirred so that no local heating occurs). The case in which a liquid is allowed to cool to its freezing point is similar. These two cases are illustrated in Figure 26.1, in which the time axis corresponds to heating or cooling respectively. These two curves correspond to a pure material in which no composition change of the solid or liquid occurs during melting. For example, during the measurement of a cooling curve for an impure liquid the composition of the liquid changes during the solidification if the solute is excluded from the crystals of solvent that form (which must be true for the present method to work). Hence, the liquid phase becomes progressively more concentrated in solute and the freezing point of the solution becomes progressively lower until freezing is complete.* The heating and cooling curves for a solution are illustrated in Figure 26.2. In this experiment the amount of solute added will be relatively small, such that the heating or melting curves will hopefully be more like Figure 26.1 than Figure 26.2, but in either case a sharp "break" in the observed curve implies the freezing point of the solvent or solution has been passed.

As is implied by the above discussion, the freezing point of a solvent or solution will be obtained by observing the heating or cooling curve. The type of apparatus used for this is illustrated in Figure 26.3. The outside test tube helps slow down the rate of heat exchange of the inner test tube with its surroundings, such that temperature changes are not so abrupt. The wire stirrer is very important as constant stirring of the solid-liquid mixture (often a "slush") is required to ensure that the thermometer is obtaining a reliable reading of the average temperature of a homogeneous mixture (e.g., if the outside portion of the solution is frozen and

---

*In some cases a *eutectic solution* of solute-solvent is formed, in which the freezing point is constant and the composition of the solid that forms is also constant.

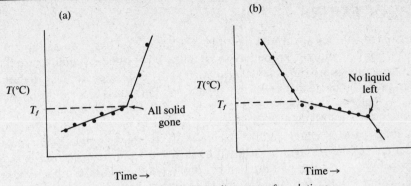

Fig. 26.2 *Heating and cooling curves for solutions.*

the interior is liquid, the temperature reading could be quite misleading). Temperature readings must be estimated to 0.2°C, and, if possible, a very accurate, sensitive thermometer should be used. However, acceptable work can be done with an ordinary −10° to 110°C laboratory thermometer. The heating or cooling curve for a particular solution should be obtained several times and the apparent freezing point averaged. A satisfactory curve can be obtained with an apparatus like that illustrated by recording the temperature at approximately 30 sec intervals.

In what follows, two choices of solvent will be considered: diphenylether (k_f = 8.00, T_f = 28°C) and naphthalene (k_f = 7.16, T_f = 80.2°C). The former material is very convenient but it tends to supercool excessively such that the heating curve must be used. The latter material has a higher freezing point, but otherwise is fairly easy to work with. Usually it is more convenient to obtain the cooling curve of naphthalene. In the following procedure it will be assumed that either diphenylether will be used with a heating curve or naphthalene will be used with a cooling curve. Your laboratory instructor may wish to assign other solvents.

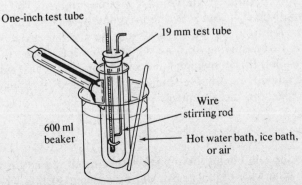

Fig. 26.3 *Apparatus for measurement of freezing point.*

II. PROCEDURE

It is assumed that an apparatus like that of Figure 26.3 has been set up and the student is familiar with the procedure to be followed to obtain a cooling curve.

1. Weigh out approximately 10 g (weighed to ±0.01 g) of the solvent and transfer to the inner test tube of the apparatus. This mass will be designated as $m_{solvent}$ in Eq. 4. Naphthalene is easy to weigh using weighing papers, but diphenylether is a little awkward to use. It is best to melt it and handle it in the liquid state (use a water bath). Because of its tendency to supercool, diphenylether can be handled as a liquid even when room temperature is several degrees below the normal melting point of diphenylether. Put the diphenylether in a small beaker (25 ml), for example, melt the diphenylether, and carefully dry the outside of the beaker. Weigh the beaker and diphenylether (to ±0.01 g), transfer approximately 10 g to the test tube, and reweigh the beaker and remaining diphenylether (to ±0.01) taking the difference to obtain the weight of diphenylether.

NOTE: *This is an example of weighing by difference (see Appendix 5).*

---

2. Obtain approximately 1 g of the unknown solute in a small, dry beaker. Weigh the beaker and solute (to ±0.002 g) and transfer 0.4–0.6 g of solute to a clean, dry test tube, and reweigh the beaker and solute (to ±0.002 g). By the same procedure, add approximately 0.2–0.4 g of solute to a second test tube. These two masses will be designated m_1 and m_2, respectively.

3. This step involves obtaining the melting point of the pure solvent. The procedures will differ for naphthalene and diphenylether.

Naphthalene

Heat a water bath to approximately 85°–90°C, and heat the inner test tube containing the naphthalene until all naphthalene has melted. Then place the outer tube of the apparatus in the water bath for a few minutes, put the tubes together as shown in Figure 26.3, and continue heating for a few more minutes. The temperature of the molten naphthalene should be approximately 83°–86°C. The test tube assembly is removed from the water bath, and the cooling curve is recorded (at ~30 sec intervals) with stirring until all naphthalene has frozen (at which point stirring is impossible). It is convenient and safe to clamp the test tube onto a ring stand during the cooldown. A curve similar to Figure 26.1(b) should be obtained. For best work, repeat this measurement.

Diphenylether

Place the inner tube containing the diphenylether in an ice bath, with the thermometer and stirring wire in place. Stir during cooling so that when solid does form you obtain a "slush" that can be stirred. As soon as solid forms remove the test tube from the ice bath. The outer test tube should be suspended in a 35°–40°C water bath (by heating gently, keep the temperature fairly constant throughout the heating curve measurement). Place the inner tube inside the outer as shown in Figure 26.3, and record the heating curve (at ~30 sec intervals), stirring constantly until all diphenylether has melted. For best work, repeat this measurement.

4. In this step, m_1 of solute (from step 2) is added to the solvent. In the case of naphthalene, the solvent will have to be remelted so the thermometer and stirrer can be removed and the solute added. For both naphthalene and diphenylether the solution must be stirred until all solute has dissolved and the solution is homogeneous. Then the cooling or heating curve should be obtained, as in step 3.

NOTE: Be sure all solute is transferred from the small test tube.

---

5. The second sample of solute m_2 is now added as above, and a new cooling or heating curve is obtained.

6. Following the graphical method shown in Figures 26.1 and 26.2, obtain the freezing point for pure solvent, designated T_f^{solvent}, solution 1 (with m_1 solute added), designated $(T_f^{\text{solution}})_1$, and solution 2 (with m_2 added), designated $(T_f^{\text{solution}})_2$.

III. DATA ANALYSIS

Calculate the freezing point depressions and apparent solute molecular weight as follows:

$$(\Delta T_f)_1 = T_f^{\text{solvent}} - (T_f^{\text{solution}})_1$$
$$(\Delta T_f)_2 = T_f^{\text{solvent}} - (T_f^{\text{solution}})_2$$

(3)

and

$$(\text{MW solute})_1 = \frac{(1000 \, k_f)}{(\Delta T_f)_1}\left(\frac{m_1}{m_{\text{solvent}}}\right)$$

$$(\text{MW solute})_2 = \frac{(1000 \, k_f)}{(\Delta T_f)_2}\left(\frac{m_1 + m_2}{m_{\text{solvent}}}\right)$$

(4)

Report both molecular weights.

IV. ERROR ANALYSIS

Estimate the error in the value of the reported molecular weight of the unknown from each of the following types of error:

1. an error in ΔT_f of $\pm 0.5°C$;

2. an error of ± 0.01 g in solvent mass;

3. an error of ± 0.002 g in m_1 and m_2 (the mass of the two samples of solute).

4. Suppose either of the solvents used was of a low purity and that the k_f constants could vary by ± 0.2 from the values stated in Table 25.1. What would the effect on the calculated molecular weight of the solute be?

5. Examine your temperature versus time graphs, and estimate the accuracy of the T_f for the pure solvent and the two mixtures. What effect will the estimated uncertainty in these individual values have on the uncertainty of the reported molecular weight?

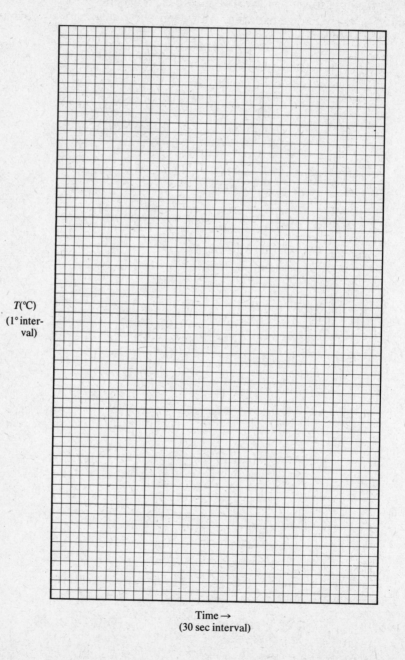

$T(^\circ C)$
(1° inter-
val)

Time →
(30 sec interval)

Fig. 26.4 *Graph for data of Experiment 26.*

NAME   DATE

SECTION

SELF-STUDY QUESTIONS

1. Very briefly describe why the addition of a solute lowers the freezing point of a material.

2. Sketch the apparatus for measuring heating or cooling curves and briefly describe the reason for the double test tube and the wire stirrer.

3. What is meant by the term *colligative property*?

4. Sketch the heating or cooling curve for a pure or impure material and indicate where the freezing point is obtained.

5. Why is the freezing or melting liquid stirred during the observation? What could happen to invalidate your data if there was no stirring?

Equilibrium and Kinetics

EXPERIMENT

27

The Ionization Constant of a Weak Acid

I. INTRODUCTION

When a weak acid (HX) dissolves in water it undergoes dissociation according to Equation 1.

$$HX \rightleftharpoons H^+ + X^- \tag{1}$$

Since the system is at equilibrium, we can write the equilibrium constant in the conventional manner (Eq. 2).

$$K = \frac{[H^+][X^-]}{[HX]} \tag{2}$$

To determine the value of the ionization constant for a weak acid, we must determine the values of $[H^+]$, $[X^-]$, and $[HX]$ in a solution containing these species, if the equilibrium is established by initially dissolving a known quantity of pure HX in a known volume of solution. Under these conditions, according to Equation 1 the following situations obtain

$$[H^+] = [X^-] \tag{3}$$

and

$$[HX] = [HX]_0 - [H^+] \tag{4}$$

where $[HX]_0$ is the molarity of the solution.

An alternative way to report on the acidity of a solution involves a quantity called pH which is defined by Equation 5.

$$p\text{H} = -\log [H^+] \tag{5}$$

Thus the pH of a solution for which the hydrogen ion concentration is 0.01 M ($[H^+] = 0.01$) is equal to 2.0. There are several ways to determine the pH of a solution experimentally.

In this experiment we will use an acid-base indicator to determine the pH of the solution in question. Indicators are themselves weak acids that have characteristic colors in both the molecular and dissolved forms. Consider the dissociation of a weak indicator acid, HIn (Eq. 6).

$$\underset{\text{color 1}}{\text{HIn}} \rightleftharpoons H^+ + \underset{\text{color 2}}{\text{In}^-} \tag{6}$$

Of course, we can write the equilibrium constant for the ionization of HIn in the usual way (Eq. 7).

$$K_{\text{HIn}} = \frac{[H^+][\text{In}^-]}{[\text{HIn}]} \tag{7}$$

When a small amount of indicator is added to an acid solution, the contribution by the ionization of HIn to the $[H^+]$ is negligible and the $[H^+]$ is fixed by the solution. Rearranging Equation 7 we obtain

$$\frac{K_{\text{HIn}}}{[H^+]} = \frac{[\text{In}^-]}{[\text{HIn}]}. \tag{8}$$

Thus the ratio of concentrations of the colored species is set by the value of the equilibrium constant of the indicator, K_{HIn}, and the $[H^+]$ in the solution. If $[H^+] \ll K_{\text{HIn}}$, then $[\text{In}^-]/[\text{HIn}]$ will be very large and the solution will appear to be color 2 (see Eq. 6). If $[H^+] \gg K_{\text{HIn}}$, then $[\text{In}^-]/[\text{HIn}]$ will be very small; the indicator will be primarily in the HIn form and the solution will appear to be color 1. If $[H^+]$ is nearly equal to K_{HIn}, both forms of the indicator will be present in about equal concentrations and the solution will appear to be a mixture of color 1 and color 2. It is difficult to detect less than about 10 percent of one form of the indicator in the presence of 90 percent of the other form; thus, most indicators change effectively from one color to the other for a change in $[H^+]$ by a factor of about 100, or about 2 pH units. It should be apparent that indicators are useful only over a relatively narrow pH range; outside of that range another indicator must be used. The colors and effective pH ranges of several common indicators appear in Table 27.1.

Given an indicator that produces a given color in its useful range, the pH of a so-

Table 27.1

Indicator	pH range	Color change
methyl violet	0–2	yellow to violet
thymol blue	1–3	red to yellow
methyl yellow	2.4–4	red to yellow
methyl red	4–6.5	red to yellow
methyl orange	3–5	red to yellow
bromphenol blue	3–5	yellow to blue
bromcresol green	4–6	yellow to blue

lution can be determined by preparing a known solution of such an [H$^+$] that the indicator has the same color in both the known and the unknown solutions.

In this experiment you will determine the pH of an unknown weak acid solution at several concentrations by means of an indicator. You will prepare solutions of known [H$^+$] to serve as references for the indicator color changes.

II. PROCEDURE

Obtain a sample of unknown acid of known molarity from the storeroom.

1. Preparation of solutions with known pH.

Draw 50 ml of 0.1 M HCl from the stock bottle into a clean erlenmeyer flask; label the flask with the pH of the solution. Using a clean pipette rinsed twice with 1–2 ml of the solution, pipette 5.0 ml of this solution into a second conical flask. Add 45 ml (graduated cylinder) of distilled water into the second flask and mix thoroughly; label this flask with its proper pH. Rinse your pipette twice with 1–2 ml of acid from this flask, and then pipette 5.0 ml of this acid into another clean flask. Dilute the acid in the third flask with 45 ml distilled water, mix thoroughly, and label it with its correct pH. Repeat this procedure until you have 5 solutions with known pH values of 1.0, 2.0, 3.0, 4.0, and 5.0.

2. The pH of the unknown acid solution.

Test 0.5 ml samples of your samples of your unknown solution with a drop of the available indicators (Table 27.1); note the colors you obtain and thus determine (roughly) the pH of your unknown solution. Then test 0.5 ml samples of your known solutions having a pH near that of your unknown with the indicator or indicators that appear to be useful in the pH range of your unknown. You should be able to determine that the pH of your unknown lies in a given one pH unit interval, e.g., between pH 3 and pH 4.

Prepare a solution with a known pH from your standard solutions so that the color of the indicator is the same as its color in your unknown. This can be done by starting with a known volume (graduated cylinder) of the more acidic standard solution with a drop of the appropriate indicator in it and adding sufficient water until the color is the same as that of your unknown solution. If the volume of this solution is significantly larger than that of your unknown, add another drop or two of indicator to this solution. Calculate the pH of the known solution; since the indicator colors match, the pH of your unknown must be the same. Be certain you look through the same depth of solutions when you make the color comparisons.

Rinse your pipette with your unknown twice, and pipette 5.0 ml of the unknown into a clean conical flask and dilute with 45 ml of distilled water. Mix the solution well and determine its pH by the method used on the original sample. Carry out two more tenfold dilutions of your unknown and determine the pH of each of these solutions. When you complete your experimental work you should have determined the pH of four solutions containing the unknown acid, the original solution, plus three successive dilutions.

III. DATA ANALYSIS

Having established the pH, and hence $[H^+]$, of your unknown solution and its 3 successive dilutions, we shall proceed to determine the equilibrium constant of the unknown acid. Rather than calculating the equilibrium constant for each dilution for which the pH was determined, we shall attempt to minimize experimental errors by using a plotting method. Substituting Equation 3 into Equation 2 we obtain

$$K = \frac{[H^+]^2}{[HA]} \qquad (9)$$

where [HA] is given by Equation 4.

Taking the logarithm of both sides of Equation 9 gives

$$\log K = 2 \log [H^+] - \log [HA]. \qquad (10)$$

Substituting Equation 5 into Equation 10 yields

$$\log K = -2pH - \log [HA] \qquad (11)$$

which upon rearrangement gives

$$2pH = - \log [HA] - \log K. \qquad (12)$$

Equation 12 is of the form $y = mx + b$ where y is identified with $2\,pH$ and x with $\log [HA]$; the slope m should be -1 and the intercept $-\log K$. Thus, a plot of the quantity $2\,pH$ of the unknown solutions against $\log [HA]$ will be a straight line with a slope of -1; the intercept will occur at $-\log K$ (when $\log [HA] = 0$). Since $\log K$ can be obtained from this graph, K can be found. Make a plot of $2\,pH$ versus $\log [HA]$ and determine K for your unknown acid.

IV. ERROR ANALYSIS

1. What is the effect on the "known" pH values if in the first dilution 4.5 ml of the $0.1\,M$ HCl was used instead of the 5.0 ml called for?

2. Consider the effect on the known $p\text{H}$ values if there is an error of ± 2 ml in the 45 ml of water used in the dilutions.

3. If the known solutions have an error of ± 0.2 $p\text{H}$ unit, will the same numerical error carry over to the $p\text{H}$ values of the unknown solutions?

4. Along the same lines as 3 above, consider the effects of ± 0.2 $p\text{H}$ unit on the plot of 2 $p\text{H}$ versus log [HA] (recall that [HA] is dependent on $p\text{H}$, see Eq. 4). Qualitatively estimate the uncertainty in log K.

5. Based on your experience in matching the colors of the known and unknown solutions, what is the experimental uncertainty in the measured pH values? How does the uncertainty compare with the uncertainty from a ± 2 ml volumetric error as in 2 above?

NAME   **DATE**

SECTION

SELF-STUDY QUESTIONS

1. What is the equilibrium constant expression for the simple ionization of the weak acid HA? What is the relationship between log K (K = equilibrium constant) and the measured pH and molarity of HA (= $[HA]_0$)?

2. How does an indicator work? If K_{HIn} (ionization constant for indicator HIn) is 10^{-4}, what is the approximate pH range this indicator can be used for?

3. If 5.0 ml of 0.15 M HCl is pipetted into a dry flask, and 40 ml of distilled water added, what is the pH of this solution? Would this calculation be performed the same way if the acid were acetic acid ($K_a = 1.8 \times 10^{-5}$) instead of HCl?

4. Describe the way in which color comparisons should be made between the color of the known solution and the unknown solution.

5. Describe the graphical method that will be used to obtain $\log K$.

EXPERIMENT

28

Colorimetric
Determination of
an Equilibrium
Constant

I. INTRODUCTION

The object of the present experiment is the determination of the equilibrium constant for the complex-ion reaction shown in Equation 1

$$Fe(H_2O)_6^{3+} + SCN^- = [Fe(H_2O)_5SCN]^{2+}. \tag{1}$$
$$\text{(colorless)} \quad \text{(colorless)} \quad \text{(red-brown)}$$

The technique applied will measure the concentration of the colored species (which will be written as $[Fe(SCN)^{2+}]$ for brevity) that results when solutions of known initial concentrations of Fe^{3+} (shorthand for $Fe(H_2O)_6^{3+}$) and SCN^{1-} are mixed. The method of analysis is colorimetry. You should review the visual colorimetric technique described in Appendix 7.

The concentration of all species present at equilibrium in a given mixture can be deduced using the following arguments.

If we define the following quantities

$$[Fe^{3+}]_{initial} = \text{known initial concentration of } Fe^{3+} \tag{2}$$

$$[SCN^-]_{initial} = \text{known initial concentration of } SCN^- \tag{3}$$

$$[Fe(SCN)^{2+}]_{final} = \textit{measured} \text{ final concentration of } Fe(SCN)^{2+} \tag{4}$$

then using Equation 1, a mass balance for the iron-containing species and the SCN^--containing species leads to Equation 5 and Equation 6

$$[Fe^{3+}]_{final} = [Fe^{3+}]_{initial} - [Fe^+(SCN)^{2+}]_{final} \tag{5}$$

$$[SCN^{1-}]_{final} = [SCN^{1-}]_{initial} - [Fe(SCN)^{2+}]_{final} \tag{6}$$

The equilibrium constant for Equation 1 can be expressed in terms of experimentally determinable quantities (Eq. 7)

$$K_{eq} = \frac{[Fe(H_2O)_5SCN^{2-}]}{[SCN^{1-}][Fe(H_2O)_6^{3+}]} = \frac{[Fe(SCN)^{2+}]_{final}}{[Fe^{3+}]_{final}[SCN^{1-}]_{final}}. \tag{7}$$

Four measurements of K_{eq} will be made and the average value reported.

II. PROCEDURE

The 7 steps that follow presume that the concentration of $Fe(SCN)^{2+}$ will be determined by visual comparison colorimetry. If a spectrophotometer is used the steps discussed in Appendix 7 will have to be carried out before the following.

1. Select 5 test tubes, 19×150 mm (if you do not have enough test tubes you may use any small beaker or other container capable of holding 10 ml of solution). Be sure they are clean and dry. Label these tubes 1–5.

2. Obtain in the lab approximately 20 ml of the stock $0.20\ M\ Fe(NO_3)_3$ solution and 30 ml of the stock $0.002\ M$ KSCN solution in reagent bottles.

3. Each test tube will contain 5 ml of a different $Fe(NO_3)_3$ solution. Using the graduated cylinder and 5 ml pipette make up 50 ml of $0.02\ M\ Fe(NO_3)_3$ solution from the $0.2\ M\ Fe(NO_3)_3$ stock solution. This latter solution will be used to make more dilute solutions of $Fe(NO_3)_3$ by means of twofold dilutions. Each test tube is to have 5 ml of the following solutions pipetted into it:

test tube 1	0.2 M	$Fe(NO_3)_3$
2	0.02 M	$Fe(NO_3)_3$
3	0.01 M	$Fe(NO_3)_3$
4	0.005 M	$Fe(NO_3)_3$
5	0.0025 M	$Fe(NO_3)_3$

The preparation of this solution is described in steps 4 through 6 below.

☛ **CAUTION:** Do not use the Fe^{3+} solution used in qualitative analysis.

---

4. To make up the $0.02\ M\ Fe^{3+}$ solution, pipette 5 ml of the stock $0.2\ M\ Fe^{3+}$ into your graduated cylinder and add distilled water until the total volume of solution is 50 ml. Mix well. Pipette 5 ml of this solution into test tube 2.

NOTE: The "golden rule" of dilutions is $c_i v_i = c_f v_f$ where c_i, v_i are the initial concentration and volume of a solution and c_f, v_f are the final concentration and volume. In the present case $c_i = 0.2\ M$, $v_i = 5\ ml$, and $v_f = 50\ ml$, so $c_f = (0.2\ M)(5\ ml)/(50\ ml) = 0.02\ M$.

---

5. To make up the $0.01\ M\ Fe^{3+}$ solution, pour out the solution prepared in 4 until 25 ml of this $0.02\ M$ solution *remain* in the graduated cylinder. Add distilled water until the total volume of solution in the graduated cylinder is 50 ml. Mix well. Pipette 5 ml of this solution into test tube 3. For this case $c_i = 0.02\ M$, $v_i = 25\ ml$, and $v_f = 50\ ml$, so

$$c_f = (0.02\ M)(25\ ml)/(50\ ml) = 0.01\ M. \qquad (8)$$

6. To make up the 0.005, $0.0025\ M\ Fe^{3+}$ solution, follow the same procedure as in step 5.

7. Into each test tube pipette 5 ml of the stock 0.002 KSCN solution. Make sure

the mixing of the solutions is complete and the color is homogeneous. Transfer the contents of test tube 1 to a 10 ml graduated test tube and label this tube as "#1."

If you are using a spectrophotometer go on to step (8b).

NOTE: The intensity comparisons should be made during the same laboratory period that the red complex is formed.

8a. The following procedure will be applied to test tubes 2–5: Transfer the contents of test tube 2 to the other 10 ml graduated test tube. Using a long medicine dropper, withdraw the solution from tube 1 until its color matches each of the other test tubes in turn (retain this solution in a clean beaker or test tube). Measure the depths of each solution to the nearest 0.5 mm (from the bottom of the test tube to the bottom of the meniscus). Record the ratios obtained in three determinations in your notebook as shown in Table 28.1.

8b. It is presumed that the absorption spectrum of $Fe(SCN)^{2+}$ has been determined and a suitable wavelength (λ_0) for monitoring the absorption has been picked. Also the molar extinction coefficient at the observation wavelength is assumed to be known or alternatively a calibration curve has been prepared (see Appendix 7). The transmission (I/I_o) or the absorbance ($\log(I/I_o)$) will be measured twice for each solution at wavelength λ_0. The data should be entered in a table similar to Table 28.1.

III. DATA ANALYSIS

The extent to which the reaction occurs can be represented mathematically by the equilibrium constant (Eq. 9)

$$K_{eq} = \frac{[Fe(SCN)^{2+}]_F}{[Fe^{3+}]_F[SCN^-]_F} \tag{9}$$

where K_{eq} is known as the equilibrium constant. In order to calculate K_{eq} you must know the final concentrations of $Fe(SCN)^{2+}$, Fe^{3+}, and SCN.

Table 28.1

Test tube	d_1/d_i			
	I	II	III	Average value
2				
3				
4				
5				

A. Analysis of Direct Visual Color Comparison Data

Determine $[Fe(SCN)^{2+}]_F$ colorimetrically by matching test tubes 2, 3, 4, and 5 to test tube 1. The $[Fe(SCN)^{2+}]_F$ in test tube 1 is known and equal to 0.001 M because this tube contains a large amount of Fe^{3+} in relationship to SCN^- which drives equilibrium 1 essentially to completion. In doing so, all the SCN^- is consumed; the maximum concentration of the red complex possible is governed by the original concentration of SCN^-.

$$\left. \begin{array}{cccc} & \dfrac{Fe^{3+}}{} & \dfrac{SCN^-}{} & \dfrac{Fe(SCN)^{2+}}{} \\ \text{Initially} & 0.2/2 = 0.1 & 0.002/2 = 0.001 & 0 \\ \text{Finally} & \text{excess} & 0 & 0.001 \end{array} \right\} \quad \text{Test Tube 1}$$

Note that since you placed 5 ml of Fe^{3+} and 5 ml of SCN^- into each test tube, the initial concentrations will be one-half of the value of the concentrations of the separate solutions, before mixing.

Calculation of $[Fe(SCN)^{2+}]_F$ in tubes 2–5 is as follows:

Tube	$[Fe(SCN)^{2+}]_F$
2	$= (0.001)(d_1/d_2)_{avg}$
3	$= (0.001)(d_1/d_3)_{avg}$
4	$= (0.001)(d_1/d_4)_{avg}$
5	$= (0.001)(d_1/d_5)_{avg}$.

B. Analysis of Spectrophotometric Data

From your table of transmissions $(I(\lambda_0)/I_0(\lambda))$ (or absorbances $(\log I_0(\lambda)/I(\lambda_0))$) and the value of $\epsilon(\lambda_0)$ (or your calibration curve) obtained from the procedure in Appendix 7, calculate the concentration of $Fe(SCN)^{2+}$ for each test tube. For example, if $T_2 = I(\lambda_0)/I_0(\lambda_0)$ for test tube 2, then

$$[Fe(SCN)^{2+}]_{\#2} = \frac{\log (1/T_2)}{\epsilon(\lambda_0) \cdot d}.$$

Nearly all the SCN^- ion reacts to form the complex. A calibration curve of $I(\lambda_0)/I_0(\lambda_0)$ (or better,* $\log (I_0(\lambda_0)/I(\lambda_0))$) versus $[Fe(SCN)^{2+}]$ can be constructed for use in the experiment.

C. Calculation of Concentration of Other Species

Calculate $[Fe^{3+}]_F$ and $[SCN^-]_F$ for each test tube using the appropriate $[Fe^{3+}]_I$, $[SCN^-]_I$, $[Fe(SCN)^{2+}]_F$ values to complete Table 28.2.

The last column, K_{eq} is calculated from Equation 16, each test tube using the appropriate values of $[Fe^{3+}]_F$, $[SCN^-]_F$, and $[Fe(SCN)^{2+}]$. You are to report K_{eq} (average) (i.e., the average of your values for tubes 2–5) and the standard deviation (Appendix 1).

---

*Better because this quantity would be expected to depend linearly on $Fe(SCN)^{2+}$ and hence interpolation between calibration points is easier.

Table 28.2

Tube	$[Fe^{3+}]_I$	$[SCN^-]_I$	$[Fe(SCN)^{2+}]_F$	$[Fe^{3+}]_F$	$[SCN]_F$	K_{eq}
2	0.01	0.001	_____	_____	_____	____
3	0.005	0.001	_____	_____	_____	____
4	0.0025	0.001	_____	_____	_____	____
5	0.00125	0.001	_____	_____	_____	____

IV. ERROR ANALYSIS

1. For several (or all) of your data points, calculate the error in $[Fe^{3+}]_F$, $[SCN^-]_F$, and $[Fe(SCN)^{2+}]_F$ that would arise from a depth error of ± 3 mm. What is the resulting uncertainty in K_{eq}?

2. Suppose an error of ± 0.5 ml were made in the original 5 ml aliquot of the 0.20 M $Fe(NO_3)_3$. What effect will this have on $[Fe^{3+}]_F$ for several of your data points? What is the possible range of K_{eq} values from this error?

3. What error in $[Fe^{3+}]_F$ can result from a ± 1 ml error in the 50 ml measurement made in carrying out the dilutions in step 3? What is the resultant error in K_{eq} for the various dilutions?

4. What is the result on $[Fe(SCN)]^{2+}$ for several or all of your data points if an error in the SCN^{-1} concentration is found to be $\pm 0.0005\ M$ (i.e., $0.0025\ M$ or $0.0015\ M$)? What will the effect be on the individual K_{eq} values?

5. Consider the standard deviation (see Appendix 1) for the average ratios recorded in Table 28.1 and consider the effect of this uncertainty on $[Fe(SCN)^{2+}]_F$ and K_{eq}.

SELF-STUDY QUESTIONS

1. How does colorimetry work and how is it applied to the present experiment?

2. How is the concentration of $Fe(SCN)^{2+}$ obtained for the different mixtures of Fe^{3+} and SCN^-? What is assumed about the initial mixture of Fe^{3+} and SCN^- (test tube 1)?

3. In comparing the color intensities of two solutions, why does one look down the length of the test tube containing the solution rather than from the side?

4. Explain what is meant by the expression $c_u d_u = c_k d_k$, where c_k, c_u are "known" and "unknown" concentrations and d_k and d_u are the depths of these 2 solutions, under certain conditions.

5. If 5 ml of $0.4\,M$ Fe^{3+} is added to a graduated cylinder and water is added until the total volume is 40 ml, what is the concentration of Fe^{3+}?

6. If for a given trial $[Fe^{3+}]_I$ is $0.005\,M$, $[SCN^-]_I$ is $0.001\,M$, and $[Fe(SCN)^{2+}]_F$ is $0.0001\,M$, what are the values of $[Fe^{3+}]_F$, $[SCN^-]_F$, and K_{eq}?

NOTE: *This is made up data and should not be compared with your experimental data!*

EXPERIMENT

29

The Temperature Dependence of the Rate of Aquation of *Trans*-Dichlorobis (ethylenediamine) Cobalt (III) Ion

I. INTRODUCTION

Experiments 9 and 29 are related. The compound synthesized in experiment 9 is to be used in this experiment to study the rate of its reaction with water—or its aquation.

The complex ion *trans*-dichlorobis(ethylenediamine)cobalt(III), the structural formula of which is shown in I, is green.

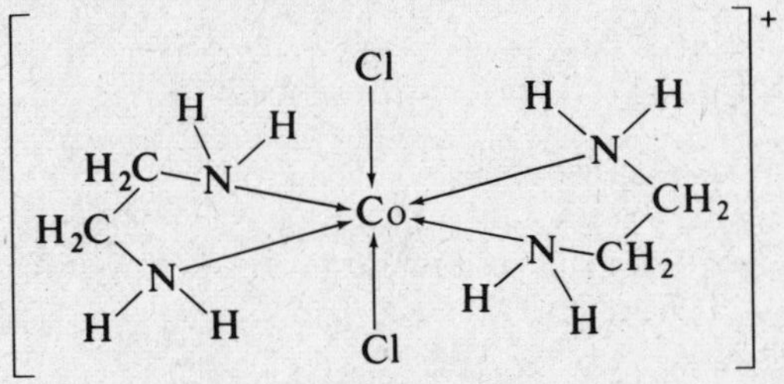

(green)

(I)

In aqueous solution one of the chloride ions is rather easily replaced by a water molecule, forming the pink ion shown as (II)

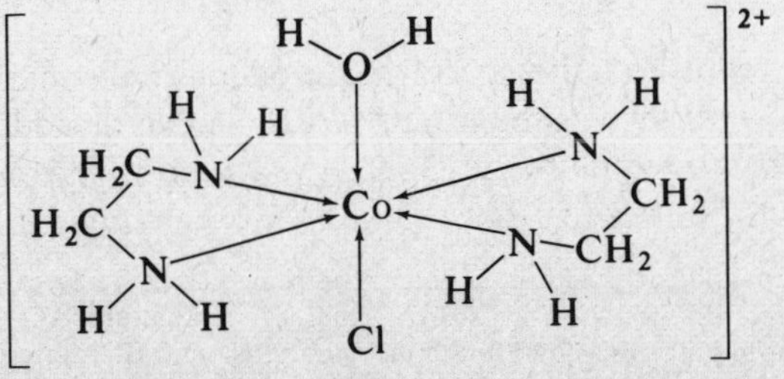

(pink)

(II)

The reaction of I with water occurs in the two steps shown in Equations 1 and 2 (*en* is the usual shorthand for the H_2N—CH_2CH_2—NH_2 molecule);

$$trans\text{-}[Co(en)_2Cl_2]^+ \xrightarrow{\text{slow}} [Co(en)_2Cl]^{2+} + Cl^- \tag{1}$$

$$[Co(en)_2Cl]^{2+} + H_2O \xrightarrow{\text{fast}} [Co(en)_2Cl(H_2O)]^{2+} \tag{2}$$

The first step (Eq. 1) corresponds to the ionization of a chloride ion from the complex I, whereas the second step (Eq. 2) involves the addition of a water molecule to the product formed in the first step.

The overall rate of aquation reaction is determined by reaction (1), which is a slow unimolecular process (i.e., a spontaneous process involving a single molecule rather than the simultaneous interaction of two or more molecules). Radioactive decay processes are also unimolecular processes. For unimolecular processes it is known that the concentration of the reactant, *trans*-$[Co(en)_2Cl_2]$ in this case, decreases with time as described by the formula shown in Equation 3

$$C_R(t) = C_Oe^{-kt} = C_O10^{-kt/2.3} \tag{3}$$

where $C_R(t)$ = concentration of reactant at time t (measured in seconds) after the reaction has begun

C_O = initial concentration of reactant before reaction started

k = rate constant for reaction (expressed in units of sec^{-1})

t = time in seconds since initiation of reaction

Qualitatively we can see from Equation 3 that the larger the value of k (which must be positive), the more rapidly the complex ion *trans*-$[Co(en)_2Cl_2]^+$ will be aquated. In practice we find that k for this reaction increases rapidly with temperature. The temperature dependence of k is predicted from *absolute reaction rate theory* to be of the form shown in Equation 4.

$$k = k_0e^{-\Delta E_f^*/RT} = k_010^{-\Delta E_f^*/2.3RT} \tag{4}$$

where k_0 = reaction rate as T approaches ∞ (i.e., $\lim_{T\to\infty} k = k_0$)

ΔE_f^* = activation energy for forward reaction (see below)

T = temperature, degrees Kelvin (°K)

R = the gas constant*

with the value of 1.9878×10^{-3} kcal/mole °K

The activation energy corresponds to an (unspecified) energy barrier that must be surmounted before the reaction can occur. The diagram in Figure 29.1 illustrates this idea.

In other words ΔE_f^* represents the energy required to remove one Cl^- ion of the complex ion which must occur before an H_2O molecule can be added to the complex. Note that ΔE_f^* has no relation to $\Delta E_{reaction}$, which is the thermodynamic energy of reaction (note, however, that the activation energy for the backward

---

*This is the same constant used for ideal gas calculations, only the numerical value is 0.08206 l-atm/mole-deg for those calculations. The units l-atm are an energy unit (l-atm represents a force × distance) and 0.08206 l-atm = 1.9878 cal. In SI units R = 8.3143 Joules/mole °K.

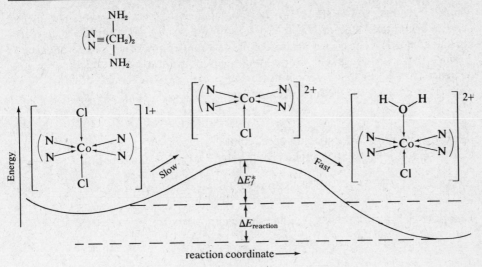

Fig. 29.1   *Potential energy diagram for* Co(en)₂Cl₂⁺ *aquation.*

reaction will be $\Delta E_b^* = \Delta E_f^* + \Delta E_{\text{reaction}}$. The reaction coordinate referred to in Figure 29.1 is a parameter that is meant to describe the progress of the reaction; it is analogous to a coordinate that describes the position of some object in space, while E is analogous to the potential energy of that object.

The primary purpose of this experiment is the study of the aquation rate k as a function of temperature. The rate at which the aquation reaction, shown in Equation 5,

$$\text{trans-}[\text{Co(en)}_2\text{Cl}_2]^{+1} + \text{H}_2\text{O} \longrightarrow [\text{Co(en)}_2\text{Cl}(\text{H}_2\text{O})]^{+2} + \text{Cl}^- \qquad (5)$$

$$\text{(green)} \qquad\qquad\qquad\qquad \text{(pink)}$$

occurs will be determined at several temperatures in the range 50–70°C by determining the length of time it requires for the originally green solution to turn *gray*. In practice the solution will be observed to pass from being faintly green–gray (at time equals $t_g$sec) to being faintly pink (at time equals $t_p$sec). The time required to reach the point where the solution contains equal amounts of reactant and product *viz.*, half of the starting material (green[Co(en)₂Cl₂]⁺) has been replaced by an equivalent amount of product (pink[Co(en)₂ClH₂O]²⁺), is measured by t_{gy}, the time for the solution to become gray. Unfortunately, the color change from green to gray to pink is not easily observed. But the time at which the solution is gray (t_{gy}) may be taken to be $t_{gy} = \frac{1}{2}(t_g + t_p)$.

When the mixture of the two complexes above is gray, the concentration of the reactant (C_R) and concentration of the product (C_P) stand in a constant ratio as given by Equation 6.

$$\frac{C_R(t_{gy})}{C_P(t_{gy})} = C_G \qquad (6)$$

where $C_G \sim 0.46$. As is shown in the following equations, however, we do not need to know the value of C_G. In Equation 6 we write $C_R(t_{gy})$ and $C_P(t_{gy})$ to emphasize that this is the concentration of the reactant and product at time equals t_{gy}.

Since there is no loss of cobalt-containing material during the reaction the sum of reactant and product concentrations must equal the concentration of the product at time $t = O$ (the time when the reaction was initiated), so we may write

$$C_R(t_{gy}) + C_P(t_{gy}) = C_O \tag{7}$$

where C_O = initial concentration of reactant.

From Equation 6 we may replace $C_P(t_{gy})$ by $C_R(t_{gy})/C_G$, so Equation 7 becomes

$$\frac{C_R(t_{gy})}{C_G} + C_R(t_{gy}) = C_O \tag{8}$$

Equation 8 is solved for $C_R(t_{gy})$ giving the result

$$C_R(t_{gy}) = C_O \frac{C_G}{C_G + 1}. \tag{9}$$

Substituting Equation 9 into Equation 3, we obtain

$$C_R(t_{gy}) = C_O \left(\frac{C_G}{C_G + 1} \right) = C_O e^{-kt_{gy}} = \frac{C_O 10^{-kt_{gy}}}{2.3}. \tag{10}$$

We may cancel the factor of C_O on both sides (which shows, incidentally, that we do not need to know the value of C_O). Taking the $\log_{10}$ of Equation 10, the relationship shown in Equation 11 is obtained.

$$\log \left(\frac{C_G}{C_G + 1} \right) = \frac{-kt_{gy}}{2.3} \tag{11}$$

or

$$\log \left(\frac{C_G + 1}{C_G} \right) = \frac{kt_{gy}}{2.3} \tag{12}$$

Solving Equation 12 for k we find

$$k = \frac{1}{t_{gy}} 2.3 \log \left(\frac{C_G + 1}{C_G} \right) \tag{13}$$

or

$$k = \frac{A}{t_{gy}} \tag{14}$$

where we have defined A as follows:

$$A = 2.3 \log \left(\frac{C_G + 1}{C_G} \right) \tag{15}$$

We observe that A is *temperature independent*.

According to Equation 4, the $\log k$ will depend on T (in °K) as follows:

$$\log(k) = \log(k_0) - \frac{\Delta E_f^*}{2.3KT} \tag{16}$$

If we take the log of Equation 14 we obtain

$$\log k = \log A + \log \frac{1}{t_{gy}} \tag{17}$$

Combine Equations 16 and 17, and rearranging we find that

$$\log \frac{1}{t_{gy}} = \log(k_0) - \log A - \frac{\Delta E_f^*}{2.3KT} \tag{18}$$

or

$$\log \frac{1}{t_{gy}} = (\text{constant}) - \left(\frac{\Delta E_f^*}{2.3K}\right) \frac{1}{T} \tag{19}$$

Equation 19 is the basis of the analysis of this experiment. It states that a plot of $\log(1/t_{gy})$ (where t_{gy} is observed at various temperatures by the student) versus $1/T$ (T = temperature in °K) should give a straight line, with a slope of $\Delta E_f^*/2.3K$. We note that the temperature dependence of a simple *unimolecular reaction*, like the one studied here, is related to ΔE_f^*: if ΔE_f^* is large, the rate decreases very rapidly with a decrease in T; if ΔE_f^* is small, the rate is not affected so adversely by a decrease in T.

In this experiment you will tabulate the value of t_{gy} for a range of temperatures between 50–70°C. Then these data will be plotted according to Equation 19 and the best straight line determined for the data points. The values of $\log(t_{gy}^{-1})$ are plotted along the Y axis and the values of T^{-1} along the X axis.

II. PROCEDURE

1. Put 5 ml of distilled H$_2$O in a 12 × 100 ml test tube and place this tube in a 250 ml beaker filled with ice. Dissolve approximately 0.5 g of the *trans*-[Co(en)$_2$Cl$_2$] − Cl in the thoroughly chilled water. Keep this solution in an ice bath throughout the experiment. Use a triple beam balance to weigh the sample.

2. While the 5 ml of H$_2$O in step 1 is being chilled, set up a bath as shown in Figure 29.2 and begin heating the water. (Do not allow the temperature to rise above 70°C.)

The bath should be arranged so that the thermometer and test tube do not touch the beaker sides or bottom and a clear view of the test tube against a white background is afforded. The thermometer should be turned so that the temperature can be read while simultaneously observing color changes in the test tube. The contents of the test tube should be stirred with a very thin stirring rod (the thinness reduces heat conduction temperature losses). The main heating bath also requires stirring to maintain good water circulation around the test tube. You should learn to control the temperature within 1°C over a 3–5 minute period by gentle

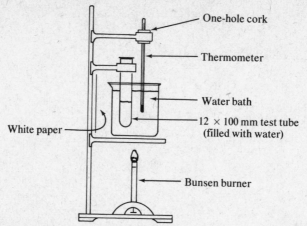

One-hole cork

Thermometer

Water bath

12 × 100 mm test tube
(filled with water)

White paper

Bunsen burner

Fig. 29.2 *Water bath and apparatus for observing aquation reaction.*

heating. During the experiment it will be necessary to maintain the bath temperature while watching for color changes.

3. Determine the height in your medicine dropper that corresponds to a volume of 0.25–0.5 ml. In the experiments that follow approximately 0.25–0.5 ml of the chilled stock *trans*-[Co(en)$_2$Cl$_2$]·Cl solution will be used for each aquation rate determination.

4. For each determination, proceed as follows:

a. The temperature of the heated bath will be stabilized at the desired temperature. The 12 × 100 mm test tube is filled up to the water bath water line only and allowed to come to temperature equilibrium with the bath for at least 5 minutes before the aquation reaction is initiated.

b. 0.25–0.5 ml of the green stock solution is added to the test tube (with rapid mixing) and the time on a sweep-second-hand clock or watch noted. While maintaining the temperature of the bath constant, the solution color will be observed. The temperature of the bath and two times will be noted and entered in the notebook: (1) the elapsed time for the solution to turn green-gray (i.e., essentially gray with a greenish tinge) and the elapsed time for the solution to turn pink-gray. These two times are referred to as t_g and t_p, respectively. These two times may be different by 10–30 sec, depending on the temperature of the reaction and the color perception of the observer.

c. After t_p has been recorded, the test tube is emptied, and filled up to the water line with distilled H$_2$O. Always leave the thin stirring rod in the test tube so that all parts of the system can thermally equilibrate. After a new temperature is established, carry out steps a to c.

The first few attempts to observe the aquation may be inaccurate until you become familiar with the nature of the color changes. The rates of aquation will vary between 10–500 sec, depending on the temperature. It is recommended that the first few runs be in the temperature range 55°–60°C. It is required that at least 6 data points be entered on the plot of t_{gy}^{-1} versus T^{-1}, and there should be at least 2 data points from each of the following temperature ranges: 52°–56°C, 57°–61°C, 62°–66°C. You should carry out 8–10 aquation rate determinations and choose the 6–8 most reliable experiments for the plot of the data.

III. DATA ANALYSIS

Prepare a table in your notebook of the following form:

T^* (°C)	T (°K)	T^{-1} (°K^{-1})	t_g^* (sec)	t_p^* (sec)	t_{gy} (sec)	t_{gy}^{-1} (sec^{-1})	$\log t_{gy}^{-1}$

The quantities marked by an asterisk (*) are directly observed. The quantity t_{gy} is calculated from the expression $t_{gy} = \frac{1}{2}(t_g + t_p)$. The data should be graphed with $\log(t_{gy}^{-1})$ plotted on the Y-axis and T^{-1} plotted on the X-axis. Determine the equation of the best straight line through the data points using the least-squares method, and draw in the line. Alternatively, plot the data on the graph supplied with this experiment and draw the best straight line through the points; calculate the slope from this plot.

From the discussion in the Introduction you can determine ΔE_f^* from your data. Using the equation for the best straight line (Appendix 2) fit to your data, you can determine the slope. According to Equation 19, the slope is identified as $\Delta E_f^*/2.3R$ or

$$\text{least-squares slope} = \frac{-\Delta E_f^*}{2.3R}. \tag{20}$$

Since the value of R is known (1.987 × 10^{-3} kcal/mol-deg), the value of ΔE_f^* may be found. If you have suitable thermodynamic tables available, compare ΔE_f^* with typical $\Delta E_{\text{reaction}}$ or $\Delta H_{\text{reaction}}$ values.

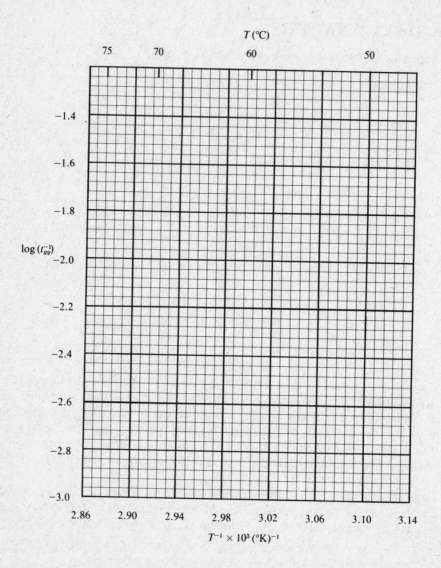

Fig. 29.3 *Graph for data, least-square line, for Experiment 29.*

IV. ERROR ANALYSIS

Estimate the uncertainty in t_{gy}^{-1} from each of the following sources of error. Estimate the maximum error in t_{gy}^{-1} from all causes and enter it in the form of error bars on your plot of t_{gy}^{-1} versus T^{-1}.

1. Everyone has a different perception of the colors green-gray or pink-gray. What error in t_{gy}^{-1} will occur if your values are always 20 sec too long or 20 sec too short? What effect will this error have on the calculated ΔE^* value?

2. Suppose that the uncertainty of t_{gy} is approximately on the order of $\pm\frac{1}{2}(t_p - t_g)(\equiv \Delta t_{gy})$. Based on your data, will this cause the percentage error in t_{gy} (i.e.,$(\Delta t_{gy}/t_{gy}) \times 100$) to be greater for large values of t_{gy} or small values of t_{gy}?

3. Based on your measured values of the dependence of t_{gy}, what will be the effect on t_{gy}^{-1} of $\pm 2°C$ error in the measured temperature?

4. Suppose that some of the $[Co(en)_2Cl_2]^+$ undergoes the aquation reaction in the ice bath. What will be the effect on the measured t_{gy} values? Would this error be more serious for t_{gy} values obtained at lower temperatures or higher temperatures? Use your value of ΔE^* to estimate the rate of aquation of $[Co(en)_2Cl_2]^+$ at 0°C. What fraction of $[Co(en)_2Cl_2]^+$ would aquate at 0°C in one hour?

NAME DATE

SECTION

SELF-STUDY QUESTIONS

1. Sketch the structure of *trans*-$Co(en)_2Cl_2^{1+}$ and $Co(en)_2Cl(H_2O)^{2+}$ ions. Which ion is pink and which is green?

2. The kinetics of the aquation reaction

$$Co(en)_2Cl_2^{1+} + H_2O \longrightarrow Co(en)_2Cl(H_2O)^{2+} + Cl^-$$

are found to be *unimolecular*. What does that term mean? How can it be applied to the above reaction?

3. The mathematical expression that often describes the temperature dependence of a reaction rate is $k = k_0 10^{-\Delta E_f^*/2.3RT}$. Identify all the symbols in this expression and give the units.

4. In this experiment the measured quantity is the time for the solution to turn gray (t_{gy}). What is the relationship of t_{gy} to the rate of reaction (k)?

5. When the original solution of $Co(en)_2Cl_2^+$ is made up it is kept in ice. What is the purpose of this?

6. The amount of $Co(en)_2Cl_2^+$ added to the warm water bath does not need to be precisely controlled? Why not?

7. Describe how the observation of the color change is supposed to be carried out, including the meaning of the symbols t_g, t_p, and t_{gy}.

EXPERIMENT

30

Kinetic Study of the Reaction of Fe^{3+} and I^-: Determination of Order with Respect to Fe^{3+}

I. INTRODUCTION

There are two complementary ways of viewing chemical reactions: thermodynamically and kinetically. Thermodynamics tells us the direction of spontaneous chemical change; in other words, it considers the stability of a chemical system and is generally restricted to systems at equilibrium. The study of kinetics, on the other hand, is concerned with how fast a chemical system can approach equilibrium, that is, the rate of the reaction. By studying the quantitative dependence of the rate on reactant concentrations, we may obtain a rate expression, or rate law, for the reaction.

This kinetic information may then be used to postulate a detailed mechanism for the reaction in which the individual steps required to convert reactants to products are specified. Kinetics also produce information for controlling the rate of a reaction. Such information is useful to the synthetic chemist, since the successful synthesis of the desired product may depend on making its rate of formation greater than the rates of competing reactions leading to other products.

In order to write the rate expression for a reaction it is necessary to determine experimentally the relationship between the rate of the reaction and the concentration of each reactant. This concentration dependence may be easily studied by varying the initial concentration of one reactant at a time while the others are held constant and by observing the effect of this change on the experimentally determined rate. Since the concentrations of the reactants decrease with time, which in turn affects the reaction rate, however, it is necessary to determine in some way the concentrations of all the reactants at the time a rate measurement is made. The method used to study the reaction in this experiment, in which variations in the initial concentration are made for one reactant while the initial concentrations of the other reactants are held constant, is called the *initial rate method*. The next step will be to measure the rate of the initial part of the reaction only so that the reactants have not had time to change appreciably from their original concentrations in the mixture. It is therefore necessary to know only the initial concentrations of the reactants in the mixture, and it is not necessary to be concerned with the presence of reaction products.

The principles of the method of initial rates may be illustrated by the hypothetical reaction represented by Equation 1.

$$2A + B \longrightarrow A_2B \tag{1}$$

The rate may be related to either the change in product concentration per unit time or the change in reactant concentration per unit time, according to Equation 2:

$$\text{rate} = -\frac{1}{2}\frac{d[A]}{dt} = -\frac{d[B]}{dt} = \frac{d[A_2B]}{dt} \tag{2}$$

where the symbol $(d[\quad]/dt)$ means the time rate of change of the species which appears in the brackets [   ]. The rate expression or rate law has the general form given in Equation 3

$$\text{rate} = k[A]^a[B]^b \tag{3}$$

where k is the specific rate constant for the reaction, and a and b represent the order of the reaction with respect to A and B, respectively. It should be emphasized that a and b are experimentally determined and are not deduced from the stoichiometry of the reaction. The order with respect to a reactant, that is, the numerical values of a and b, may be positive or negative, integral or fractional, including zero.

The top curve of Figure 30.1 shows the behavior with time of a system containing initially only A and B. The rate of the forward reaction is large at the start, but decreases as predicted from Equation 3 as the reactants are converted to A_2B. If the reaction is reversible, as assumed in Figure 30.1, the rate of the reverse reaction is related to the concentration of A_2B. Since A_2B is not present initially, the rate of the reverse reaction is zero at the start, but as A_2B is formed by the forward reaction the rate of the reverse reaction gradually increases. Eventually, of course, the forward and reverse rates become equal, and the system is then at equilibrium.

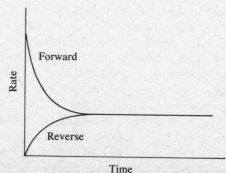

Fig. 30.1   *Change with time of the forward and reverse reaction rates for a reversible reaction.*

We assumed in Equation 2 that the rate of formation of product is related to the forward reaction rate according to the expression

$$\frac{d[A_2B]}{dt} = \text{rate}_{\text{(forward)}}. \tag{4}$$

If the reaction is reversible, however, the correct expression is given by Equation 5 which indicates

$$\frac{d[A_2B]}{dt} = \text{rate}_{\text{(forward)}} - \text{rate}_{\text{(reverse)}} \tag{5}$$

that the time rate of change of A_2B depends upon the rate it is formed ($\text{rate}_{\text{(forward)}}$) and the rate it is lost ($\text{rate}_{\text{(reverse)}}$). This observation greatly complicates the calculations, and we generally try to work under conditions where the reaction (1) is not reversible or (2) the reverse rate can be set equal to zero. This condition is nearly valid at the very start of the reaction. An obvious advantage of the initial rate method is that the reverse reaction can be ignored, since we observe only the first part of the reaction, where the reverse rate is virtually zero.

The reaction to be studied in this experiment is the oxidation of I^- by Fe^{3+} ions according to Equation 6.

$$2Fe^{3+} + 3I^- \longrightarrow 2Fe^{2+} + I_3^- \tag{6}$$

The expected rate expression for this reaction is given in Equation 7, where

$$\text{rate} = -\frac{1}{2}\frac{d[Fe^{3+}]}{dt} = \frac{d[I_3^-]}{dt} = k[Fe^{3+}]^a[I^-]^b \tag{7}$$

the values of a and b must be determined experimentally. In this experiment we will be concerned with determining the value of the a exponent. In Experiment 31, b will be determined by similar methods.

The initial rate is determined by measuring the time in seconds required for part (about 4×10^{-5} mole) of the Fe^{3+} to be reduced to Fe^{2+}. This is indicated by adding starch solution and a small, constant amount of $S_2O_3^{2-}$ to each mixture. The following reactions then occur:

$$2Fe^{3+} + 3I^- \longrightarrow 2Fe^{2+} + I_3^- \tag{8}$$

$$I_3^- + 2S_2O_3^{2-} \xrightarrow{\text{fast}} 3I^- + S_4O_6^{2-} \tag{9}$$

As soon as the $S_2O_3^{2-}$ has been consumed, any additional I_3^- formed by the reaction between ferric and iodide ions will react with the starch to form a characteristic blue color (see Fig. 30.2.)* Note that when the blue color first appears, the decrease in the concentration of Fe^{3+} from its initial value is just equal to the initial concentration of $S_2O_3^{2-}$ in the mixture. Thus the initial rate, $-\frac{1}{2}d[Fe^{3+}]/dt$, is equal to $\frac{1}{2}[S_2O_3^{2-}]_i/\Delta t$, where $[S_2O_3^{2-}]_i$ is the initial concentration of $S_2O_3^{2-}$ and Δt is the time in seconds between mixing and the appearance of the blue color. In order to

---

*Note that for this method to work, reaction 9 must be significantly faster than reaction 8, and $S_2O_3^{2-}$ must not interfere with the $Fe^{3+} + 3I^-$ reaction. These conditions are met for this reaction.

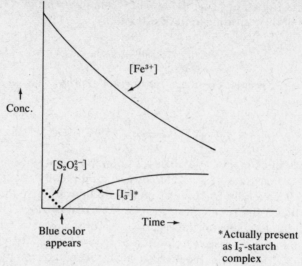

Fig. 30.2 *Schematic representation of changes in concentration in various species in this experiment.*

obtain reasonable reaction times, it is necessary to use initial rate intervals which allow the Fe^{3+} concentration to decrease slightly (about 4 to 10 percent) from its initial value. To compensate for this change, it is suggested that the average Fe^{3+} concentration during this time interval be used in place of the initial concentration.

Ions other than ferric or iodide may affect the rate by (1) varying the activity of ions in solution (the "ionic strength"), (2) serving as catalysts, or (3) serving as inhibitors. Changing the ionic strength affects the activity of the ions reacting and therefore changes the rate. This effect will not be studied. Catalysts provide a new and easier path for the reaction and thereby increase the rate. Inhibitors decrease the rate by lowering the concentration or the activity of some species, which is important in the reaction mechanism.

The effect of temperature will be discussed only briefly here because this subject is covered in detail in the discussion of Experiment 29. Recall that the specific rate constant k is related to the absolute temperature T by Equation 10

$$k = Ae^{-E_a/RT} = A\ 10^{-E_a/2.3RT} \tag{10}$$

where E_a is the activation energy for the reaction, that is the minimum energy required for the reactant to pass over an energy barrier to form products. Such information is important for a comprehensive kinetic picture of a chemical system, but it will not be covered here because the subject of activation energy has been covered in depth in Experiment 29. We should note, however, that the temperature dependence of the rate constant implies that all measurements in this experiment must be made at a constant temperature in order to obtain accurate data.

II. PROCEDURE

A. Materials and Solutions

Obtain the following materials: (1) constant temperature water bath (if available); (2) thermometer, accurate to $\pm 0.1°C$; (3) stop watch; (4) a total of three burets are required (graduated cylinders can be used but with a loss of accuracy); (5) two 10 ml pipettes; (6) 0.04 M KI solution (about 100 ml); (7) 0.004 M $S_2O_3^{2-}$ solution (about 90 ml); and (8) 2 percent starch solution (about 50 ml).

The following solutions must be prepared:

1. Ferric nitrate: This solution must be 0.04 M in $Fe(NO_3)_3$ and 0.15 M in HNO_3 to prevent hydrolysis of Fe^{3+} ion. Combine 100 ml of 0.2 M Fe^{3+} stock solution (in the laboratory) with 13 ml of 6 M HNO_3 and dilute this mixture to 500 ml in a volumetric flask (half the measurements for a 250 ml volumetric flask).

2. Nitric acid: Dilute 13 ml of 6 M HNO_3 to 500 ml to form a 0.15 M HNO_3 solution.

B. Determination of the Reaction Order with Respect to Fe^{3+}

1. Fill the three burets with H_2O, $Fe(NO_3)_3$, and HNO_3 solutions. (If sufficient burets are not available graduated cylinders may be used.) You will use the 10 ml pipettes to dispense the KI and $S_2O_3^{2-}$ solutions. You may use a 5 ml pipette to dispense the starch solution, although high accuracy is not required for the amount of starch solution transferred.

2. Since the reactions 8 and 9 begin as soon as ferric ions are in the presence of iodide ions, we will prepare two solutions that will be poured together when we are ready to time the reaction. In five erlenmeyer flasks (250 ml or 500 ml will do), prepare the following solutions:

Flask No.	0.04 M Fe^{3+}, ml	0.15 M HNO$_3$, ml	H$_2$O, ml
1	10.00	20.00	20.00
2	15.00	15.00	20.00
3	20.00	10.00	20.00
4	25.00	5.00	20.00
5	30.00	0.00	20.00

Number five beakers having a capacity of 100 ml or more. Into each add the following solutions: 10.0 ml of 0.04 M KI; 10.0 ml of 0.004 M $S_2O_3^{2-}$; 5.0 ml starch; and 25.0 ml H_2O.

NOTE: If this solution is blue before the Fe^{3+} solution is added, the solutions have been contaminated and have to be remade.

---

3. Allow the solutions to come to constant temperatures. (Place all flasks and

beakers in the water bath for 10–15 minutes; the water bath should be set for room temperature.)* Measure and record the initial temperature of flask 1.

4. Add beaker 1 to flask 1 *rapidly* and begin timing the reaction *immediately*. You may want to remove the flask and beaker from the water bath momentarily to do your pouring; swirl the flask once or twice to aid in mixing.

5. At the first appearance of blue color stop the stop watch and take the final temperature. Record Δt and the average temperature.

6. Repeat steps 4 and 5, add beaker 2 to flask 2, and so on. Tabulate the temperature and Δt values in the form below:

Trial number	T_{avg}	Δt	Initial rate	Log rate	$[Fe^{3+}]_i$	$[Fe^{3+}]_{avg}$	$\log[Fe^{3+}]_{avg}$
1							
2							
3							
4							
5							

III. DATA ANALYSIS

A. Initial Rate Calculation

The initial rate of the reaction is given by

$$rate = \frac{[S_2O_3^{2-}]_i}{2\Delta t}$$

where $[S_2O_3^{2-}]_i$ is the initial concentration of $S_2O_3^{2-}$ and Δt is the reaction time from mixing until the appearance of blue color. For a reaction that contained 10.00 ml of 0.004 M $S_2O_3^{2-}$ in a total volume of 100.00 ml and took 40 seconds to react, we have

$$rate = \frac{[(0.004\ M)(10.00\ ml)]/100\ ml}{2 \times 40\ sec} = 5.0 \times 10^{-6}\ mole\ sec^{-1}.$$

B. $[Fe^{3+}]_{avg}$ Calculation

In calculating $[Fe^{3+}]_{avg}$, remember that the blue color will first appear when all $S_2O_3^{2-}$ has been consumed:

$$[Fe^{3+}]_{final} = [Fe^{3+}]_i - [S_2O_3^{2-}]_i.$$

Therefore for $[Fe^{3+}]_{avg}$, we take the iron concentration present when *half* the $[S_2O_3^{2-}]_i$ has been consumed:

---

*Equilibrium in air will be satisfactory. The reaction in step 4 should be carried out with the flask partially submerged in a beaker of room temperature water to hold the solution temperature constant.

$$[Fe^{3+}]_{avg} = [Fe^{3+}]_i - \tfrac{1}{2}[S_2O_3^{2-}]_i.$$

For example, in Trial 1, we find

$$[Fe^{3+}]_{avg} = (4 \times 10^{-3}M) - (2 \times 10^{-4}M) = 3.8 \times 10^{-3}M.$$

Note that $[I^-]$ is constant so long as there is $S_2O_3^{2-}$ present, according to reaction 9.

Using the data from trials 1 through 5, plot log (initial rate) versus log $[Fe^{3+}]_{avg}$. Draw the best straight line (use a least-squares fit, Appendix 2) through the experimental points and determine a, the reaction order for Fe^{3+}. From Equation 7 we see that log (rate) $= a$ log $[Fe^{3+}]_{avg} + b$ log $[I^-]$ (where $[I^-]$ is constant for this case). Hence the slope of the log-log plot yields a.

IV. ERROR ANALYSIS

1. Estimate the error in the value of the initial rate that could arise from the following:

   a. errors in Δt of ± 2 sec for several trials;

   b. errors in the initial concentration of $S_2O_3^{2-}$ of $\pm 0.001\, M$;

   c. an error of ± 1 ml in the volume of $0.004\, M$ $S_2O_3^{2-}$ solution added;

   d. an error of ± 5 ml in the total volume of the solutions that are combined to start the reaction (e.g., the amount of H_2O added was wrong).

2. Estimate the error in $[Fe^{3+}]_{avg}$ for each trial that can arise from an error of ± 0.5 ml in the volume of $0.04\, M$ Fe^{3+} added to flasks 1 through 5.

3. Consider which of the possible experimental errors results in the greatest uncertainty in the initial rate values, and estimate the value of this uncertainty for each point plotted in the plot of log (rate) versus log $[Fe^{3+}]_{avg}$. Show this uncertainty in the form of error bars on your plot [parallel to the log (initial rate) axis]. Based on an estimate like question 2 above, what is the likely uncertainty in $[Fe^{3+}]_{avg}$? Show this uncertainty in the plot as an error bar (parallel to the log $[Fe^{3+}]$ axis). Compare your least-squares value of a with the nearest integral value of a by plotting both straight lines. Do both straight lines generally pass through the error bars you have estimated?

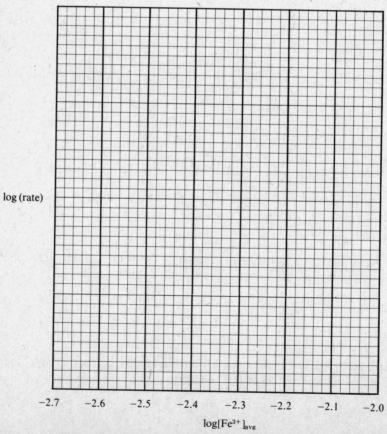

Fig. 30.3 *Graph for data, least-square line, for Experiment 30.*

SELF-STUDY QUESTIONS

1. What is meant by a rate law?

2. Describe briefly what the principles are for the method of initial rates.

3. The value of Δt (time required for the blue I_3^--starch complex to be formed) will depend *critically* on the number of moles of $S_2O_3^{2-}$ added to the reaction mixture. Why is this?

4. Why is the exact amount of 2 percent starch solution added not particularly important to the accuracy of this experiment?

5. At what point does one begin to measure the elapsed time in this experiment?

6. In this experiment $[Fe^{3+}]_i$ is varied and $[I^-]_i$ is held constant. Why is this done? Why is it inconvenient to vary both concentrations simultaneously?

7. Suppose the initial concentration of $S_2O_3^{2-}$ in the reacting solution were 2.0×10^{-4} M and the elapsed time (Δt) were 25 sec. What is the initial rate?

8. If you know the rate law for this reaction (i.e., the values of a and b), how can you obtain the rate constant (k in Equation 7)?

EXPERIMENT

31

Kinetic Study of the Reaction of Fe^{3+} and I^-: Determination of Order with Respect to I^-

I. INTRODUCTION*

This experiment is a continuation of Experiment 30, and if time permits both experiments should be done. The reaction between Fe^{3+} and I^- is as follows:

$$2Fe^{3+} + 3I^- \longrightarrow 2Fe^{2+} + I_3^- \tag{1}$$

In the presence of a small amount of thiosulfate ($S_2O_3^{2-}$) the I_3^- quickly reacts,

$$I_3^- + 2S_2O_3^{2-} \xrightarrow{\text{fast}} 3I^- + S_4O_6^{2-} \tag{2}$$

When the $S_2O_3^{2-}$ is exhausted I_3^- is formed, and in the presence of starch, a dark blue complex is formed. The time for the blue color to develop (Δt) will be measured for various initial I^- concentrations and fixed Fe^{3+} concentrations. The initial rate of the reaction is given by $\frac{1}{2}[S_2O_3^{2-}]_i/\Delta t$ where $[S_2O_3^{2-}]_i$ is the concentration of S_2O_3 at the beginning of the reaction. The rate law for this reaction may be written

$$\text{initial rate} = \frac{1}{2}\frac{[S_2O_3^{2-}]_i}{\Delta t} = k[Fe^{3+}]_{avg}^a[I^-]_i^b$$

where

$$[Fe^{3+}]_{avg} = [Fe^{3+}]_i - \frac{1}{2}[S_2O_3^{2-}]_i$$

($[Fe^{3+}]_i$ is the initial concentration of Fe^{3+}, 0.004 M for this experiment). By plotting log (initial rate) versus log $[I^-]_i$, for constant $[Fe^{3+}]_{avg}$, the exponent b can be determined.

---

*The Introduction to Experiment 30 should be read before doing this experiment.

II. PROCEDURE

A. Materials and Solutions

The materials and solutions for this experiment are the same as those for Experiment 30, which we repeat here for convenience. Obtain the following materials: (1) constant temperature water bath (if available); (2) thermometer, accurate to $\pm0.1°C$; (3) stop watch; (4) a total of three burets are required (graduated cylinders can be used but with a loss of accuracy); (5) two 10 ml pipettes; (6) 0.04 M KI solution (about 100 ml); (7) 0.004 M $S_2O_3^{2-}$ solution (about 90 ml); and (8) 2 percent starch solution (about 50 ml).

The following solutions must be prepared:

1. Ferric nitrate: This solution must be 0.04 M in $Fe(NO_3)_3$ and 0.15 M in HNO_3 to prevent hydrolysis of Fe^{3+} ion. Combine 100 ml of 0.2 M Fe^{3+} stock solution (in the laboratory) with 13 ml of 6 M HNO_3, and dilute this mixture to 500 ml in a volumetric flask.

2. Nitric acid: Dilute 13 ml of 6 M HNO_3 to 500 ml to form a 0.15 M HNO_3 solution.

B. Determination of the Reaction Order with Respect to I^-

This portion of the experiment parallels part B in section II of Experiment 30, but here the concentration of I^- varies.

1. Number five flasks 1–5. The following mixture will be added to each flask: 10.00 ml 0.04 M Fe^{3+}; 20.00 ml 0.15 M HNO_3; 20.00 ml H_2O.

2. Rinse a buret several times with water and once with KI solution, then fill with KI solution. Prepare the following beakers:

Beaker No.	0.04 M KI, ml	0.004 M $S_2O_3^{2-}$, ml	Starch, ml	H_2O, ml
1	5.00	10.00	5.00	30.00
2	10.00	10.00	5.00	25.00
3	15.00	10.00	5.00	20.00
4	20.00	10.00	5.00	15.00
5	25.00	10.00	5.00	10.00

3. Allow the solutions to come to constant temperatures. (Place all flasks and beakers in the water bath for 10–15 minutes; the water bath should be set for room temperature.)* Measure and record the initial temperature of flask 1.

4. Add beaker 1 to flask 1 *rapidly* and begin timing the reaction *immediately*. You may want to remove the flask and beaker from the water bath momentarily to do your pouring; swirl the flask once or twice to aid in mixing.

---

*Equilibrium in air will be satisfactory. The reaction in step 4 should be carried out with the flask partially submerged in a beaker of room temperature water to hold the solution temperature constant.

5. At the *first* appearance of blue color stop the stop watch and take the final temperature. Record Δt and the average temperature.

6. Repeat steps 4 and 5, add beaker 2 to flask 2, and so on. Tabulate the temperature and Δt values in the form below:

Trial number	T_{avg}	Δt	Initial rate	Log (initial rate)	$[I^-]$	$Log[I^-]$
1						
2						
3						
4						
5						

III. DATA ANALYSIS

As discussed in Experiment 30, the initial rate is given by

$$\text{initial rate} = \frac{[S_2O_3^{2-}]_i}{2\Delta t} = \frac{(0.004\ M)(10\ \text{ml})}{(100\ \text{ml})} \frac{1}{2\Delta t}.$$

Since $[Fe^{3+}]_i$ is constant in this experiment, it is not necessary to compute $[Fe^{3+}]_{avg}$ because the amount of Fe^{3+} consumed will be the same for each trial. As can be seen from Equation 2, $[I^-]$ is constant so long as $S_2O_3^{2-}$ is present, so it is not necessary to compute $[I^-]_{avg}$ (i.e., $[I^-]_{avg} = [I^-]_i$).

Using the data from trials 1 through 5, plot log (initial rate) versus log $[I^-]$. Obtain the best straight line through the data using a least-squares fit (see Appendix 2), and from the slope determine b.

IV. ERROR ANALYSIS

1. Estimate the error in the value of the initial rate that could arise from the following:

a. errors in Δt of ± 2 sec for several trials;

b. errors in the initial concentration of $S_2O_3^{2-}$ of $\pm 0.001\,M$;

c. an error of ± 1 ml in the volume of $0.004\,M$ $S_2O_3^{2-}$ solution added;

d. an error of ± 5 ml in the total volume of the solutions that are combined to start the reaction (e.g., the amount of H_2O added was wrong).

2. Estimate the error in $[I^-]$ for each trial that can arise from an error of ± 0.5 ml of $0.04\,M$ I^- added to flasks 1 through 5.

3. Consider which of the possible experimental errors results in the greatest uncertainty in the initial rate values, and estimate the value of this uncertainty for each point plotted in the plot of log (rate) versus log $[I^-]$. Show this uncertainty in the form of error bars on your plot (parallel to the log (initial rate)). Based on an estimate like question 2 above, what is the likely uncertainty in $[I^-]$? Show this uncertainty in the plot as an error bar (parallel to the log $[I^-]$ axis). Compare your least-squares value of b with the nearest integral value of b by plotting both straight lines. Do both straight lines generally pass through the error bars you have estimated?

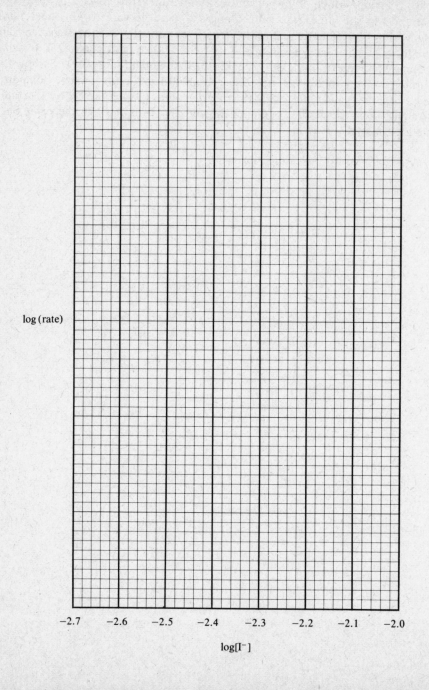

log (rate)

−2.7   −2.6   −2.5   −2.4   −2.3   −2.2   −2.1   −2.0

log[I⁻]

Fig. 31.1 *Graph for data, least-square line, for Experiment 31.*

NAME _____   **DATE** _____

SECTION _____

SELF-STUDY QUESTIONS

1. What is meant by a rate law?

2. Describe briefly what the principles are for the method of initial rates.

3. The value of Δt (time required for the blue I_3^--starch complex to be formed) will depend critically on the number of moles of $S_2O_3^{2-}$ added to the reaction mixture. Why is this?

4. Why is the exact amount of 2 percent starch solution added not particularly important to the accuracy of this experiment?

5. At what point does one begin measuring the elapsed time in this experiment?

6. Sketch the way in which the concentrations of Fe^{3+}, I^-, I_3^-, and $S_2O_3^{2-}$ vary with time. At what point does the blue color appear?

7. If the initial concentration of $S_2O_3^{2-}$ is 3.0×10^{-4} M, and the time elapsed before the blue color appears is 20 sec, what is the initial rate? Suppose the concentration of Fe^{3+} or I^- remains constant but the concentration of $S_2O_3^{2-}$ is changed to 2.0×10^{-4} M. What value of Δt is expected?

Thermochemistry, Thermodynamics

EXPERIMENT

32

Determination of ΔH of Neutralization for NH_3 + HX

I. INTRODUCTION

One of the most useful quantities from thermodynamics is the state function *enthalpy*, symbolized by H, and changes in enthalpy, ΔH. This latter quantity is measured by the heat evolved or absorbed for a constant pressure process. Measurements of changes in heat require a device known as a *calorimeter*, which is nothing more than a vessel insulated to prevent loss of heat to the surroundings and equipped with temperature measuring instrumentation. For the present experiment the calorimeter will be approximated by a styrofoam cup and a fairly sensitive thermometer. In all cases the amount of heat evolved or absorbed by the process is given by $\Delta Q = C_P \Delta T$, where ΔQ is the change of heat accompanying the process, C_P is the total heat capacity of the calorimeter (at constant pressure) and its contents, and ΔT is the change of temperature of the calorimeter. ΔT can be either positive or negative according to whether the process is exothermic or endothermic (e.g., ΔH is negative or positive). In the present case two reagents,

NH$_3$ (aq) and HX (aq) (some acid), will be mixed and ΔT for the reaction measured. Using the approximate heat capacity of the mixture and the measured heat capacity of the calorimeter (the so-called "calorimeter constant"), ΔQ for the reaction will be obtained, and hence ΔH for the reaction. While this experiment is simple in concept, considerable care is required to obtain reproducible, meaningful data. Small errors in measuring temperatures or allowing solutions to equilibrate before mixing can lead to very large errors.

There are two points of technique that require additional comment, the "calorimeter constant" (the heat capacity of the calorimeter itself) and the "graphical method" used to obtain ΔT for a process.

1. Calorimeter Constant

The total change of heat caused by the reaction NH$_3$(aq) + HX(aq) is given by

$$\Delta Q = (C_P)_{total}(\Delta T)_{measured} = [(C_P)_{sol} + (C_P)_{cal}] (\Delta T)_{measured}$$

where $(C_P)_{sol}$ = heat capacity of the solution after reaction, estimated as 0.95 cal ml^{-1} deg^{-1} × (volume of solution in ml)

$(C_P)_{cal}$ = heat capacity of styrofoam cup and thermometer, to be determined experimentally, and expected to be in the range of 1.0 to 10.0 cal deg^{-1}

The heat capacity of the calorimeter is determined by measuring ΔT for a known reaction, NaOH(aq) + HCl(aq) = Na$^+$(aq) + Cl$^-$(aq) + H$_2$O, and using the known ΔH of this reaction, $-13,600$ cal mole^{-1}, to compute $(C_P)_{cal}$ from the following expression:

$$(\Delta Q)_{std} = [(C_P)_{sol}^{NaCl} + (C_P)_{cal}] (\Delta T)_{meas}^{std}$$

where $(\Delta Q)_{std}$ = heat evolved by standard (std) reaction above
= 13,600 cal mole^{-1} × n_{H_2O}

where n_{H_2O} = number of moles of H$_2$O produced by the reaction, given by the normality of the HCl solution times the number of *liters* of HCl solution used (see procedure)

$(C_P)_{sol}^{NaCl}$ = heat capacity of the NaCl solution that results from the standard reaction, taken to be 0.97 cal ml^{-1} deg^{-1} × (volume of solution in ml)

$(\Delta T)_{meas}^{std}$ = measured change of temperature that results from the standard reaction; for best work this should be the average of two or three determinations

Given these quantities the equation above can be solved for $(C_P)_{cal}$, i.e.,

$$(C_P)_{cal} = \frac{(\Delta Q)_{std}}{(\Delta T)_{meas}^{std}} - (C_P)_{sol}^{NaCl}$$

NOTE: If one calculates a negative *$(C_P)_{cal}$ from the above something has gone seriously wrong. If the calculated $(C_P)_{cal}$ is negative do not proceed with the experiment until the problem is cleared up.*

---

2. Graphical Method for ΔT

At first glance the measurement of ΔT would seem straightforward. However, certain time dependent features enter into the temperature values. First, the experimentalist (you) cannot simultaneously measure temperatures and mix together reagents, so a measurement of the temperature is made just before and just after the initiation of the reaction. Second, a finite amount of time is required for temperature equilibrium of the calorimeter walls, thermometer, and solution. Third, the temperature of the solution after reaction will be changed from room temperature and the solution will cool (if the reaction is exothermic where ΔT is positive, or vice versa for an endothermic reaction). This state of affairs is illustrated in Figure 32.1, where the experiment points are indicated by solid dots, and the graphical method used to estimate ΔT is also illustrated. Each experiment requires a separate plot of temperature vs. time as in Figure 32.1, with temperatures recorded at approximately 30 second intervals.

II. PROCEDURE

Determination of Calorimeter Constant

1. Arrange the styrofoam cup and thermometer as in Figure 32.2. There should be adequate clearance between the bottom of the cup and the bottom of the thermometer to allow the solution to be swirled easily, but the bulb of the thermometer must be completely immersed when 100 ml of liquid is placed in the cup (try using 100 ml of water when setting up the arrangement for the first time).

2. Using a graduated cylinder or buret, measure 50 ml of standard 2.00 N HCl and pour into the calorimeter and measure the temperature for several minutes before carrying out step 3.

3. Rinse out the graduated cylinder with distilled water, drain until no more than a few drops of water remain, then measure 50 ml of 2.05 N NaOH (a slight

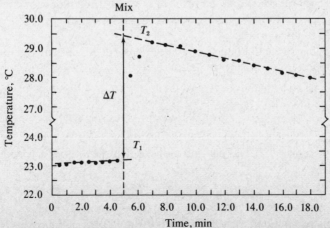

Fig. 32.1 *Typical temperature vs. time graph, illustrating the method of obtaining ΔT.*

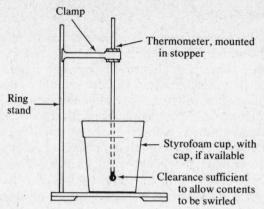

Fig. 32.2 *Experimental set-up of styrofoam cup calorimeter.*

excess of base is used to ensure a complete consumption of the acid). *It is presumed that the* NaOH *and* HCl *solutions are at the same temperature.* If there is any doubt about this, measure the temperature of the NaOH solution also, taking care to wash the thermometer off with water before putting into NaOH from the HCl solution or vice versa (you don't want any premature HCl, NaOH reaction). After it is clear that both solutions are at the same temperature, the NaOH solution may be added to the calorimeter (quickly, but without splashing).

NOTE: Temperature measurements should be taken and recorded just before and after mixing, and every 30 seconds thereafter. Swirl the contents of the calorimeter gently between temperature readings to ensure mixing and a homogeneous solution.

---

4. After all temperature data have been taken, discard the solution, rinse the calorimeter and thermometer several times, and let drain until nearly dry. This portion of the experiment should be repeated two or three times to improve the accuracy of $(C_P)_{cal}$ and to give the experimenter some experience with the general reproducibility of the method.

5. Before going on, calculate the average calorimeter heat capacity using the following equation for each determination:

$$(C_P)_{cal} = \frac{(\Delta Q)_{std}}{(\Delta T)^{std}_{meas}} - (C_P)^{NaCl}_{sol}$$

where $(\Delta Q)_{std} = 13,600 \text{ cal mole}^{-1} \times 0.100 \text{ mole} = 1360 \text{ cal}$
(note that 0.100 mole of HCl is reacting with 0.100 mole of NaOH, producing 0.100 mole of NaCl (aq))

$$(C_P)^{NaCl}_{sol} = 0.97 \text{ cal ml}^{-1} \text{ deg}^{-1} \times 100 \text{ ml} = 97 \text{ cal deg}^{-1}.$$

If $(C_P)_{cal}$ seems reasonable, then proceed to the next part of the experiment.

Determination of $\Delta H_{\text{neutralization}}$ **for NH₃ (aq) and Unknown Acid**

1. Obtain sufficient 2.05 *N* NH₃(aq) and 2.00 *N* unknown acid HX to carry out two or three ΔT determinations, as time permits. Be sure both solutions are equilibrated to room temperature. The NH₃(aq) solution should be in a closed vessel to prevent loss of NH₃(gas), which also tends to cool the solution.

2. Carry out several determinations of ΔT for the reaction of 50 ml of 2.05 *N* NH₃(aq) and 50 ml of 2.00 *N* HX following the same procedure as used in determining $(C_P)_{\text{cal}}$. Always rinse out all glassware and the styrofoam cup and drain well between measurements.

III. DATA ANALYSIS

$\Delta H_{\text{neutralization}}$ Calculation

Calculate ΔQ for the reaction NH₃(aq) + HX(aq) = NH₄X(aq) for each measurement as follows:

$$\Delta Q = [(C_P)_{\text{cal}} + (C_P)_{\text{sol}}](\Delta T)_{\text{meas}}$$

where $(C_P)_{\text{cal}}$ = quantity determined in previous section
   $(C_P)_{\text{sol}}$ = 0.95 cal ml^{-1} deg^{-1} × 100 ml
       = 95 cal deg, an estimate based on typical solution.

Calculate $\Delta H_{\text{neutralization}}$ as follows:

$$\Delta H_{\text{neutralization}} = -\frac{\langle \Delta Q \rangle}{0.100}\frac{\text{cal}}{\text{mole}} \pm \frac{\sigma}{0.100}\frac{\text{cal}}{\text{mole}}$$

where this reaction produced 0.100 mole of NH₄X(aq), and the sign of $\Delta H_{\text{neutralization}}$ follows the convention for ΔH of an exothermic reaction. $\langle \Delta Q \rangle$ is the average value of ΔQ for the various experiments, and σ is the standard deviation of $\langle \Delta Q \rangle$ (see Appendix 1).

IV. ERROR ANALYSIS

Calculate the maximum uncertainty in $\Delta H_{\text{neutralization}}$ that arises from the following sources:

1. Temperature reading errors of ±0.2°C.

2. A difference of the NH_3(aq) and HX (aq) solution temperatures of $\pm 2°C$ before mixing. This would affect the initial temperature of the reaction mixture by approximately $\pm 1°C$. Note that the error caused by this effect is computed the same way as for 1 above.

3. Errors in the volumes of combining reagents by ± 1 ml, from inaccuracy of the glassware or in reading the volume graduations. You will have to use your calculated $\Delta H_{neutralization}$ to estimate this effect.

4. Errors in the stated normality of the acid of ± 2 percent. This calculation is similar to 3 above.

5. Inaccuracies in the stated heat capacity of the solution per ml on the order of ± 0.02 cal ml^{-1} deg^{-1}.

6. Accidently leaving 5 ml of water (heat capacity approximately 5 cal deg⁻¹) in the calorimeter in addition to the added volume of reagent (i.e., this could arise from failing to drain the glassware properly).

7. Based on the above calculations, or any other errors not considered above, estimate a reasonable experimental error for $\Delta H_{\text{neutralization}}$. Compare this value with σ. If there is a serious discrepancy between the two values, try to analyze the reason(s).

NAME   **DATE**

SECTION

SELF-STUDY QUESTIONS

1. What is a calorimeter?

2. If ΔT for a reaction is negative is the reaction endothermic or exothermic?

3. What is the mathematical relationship between ΔQ, ΔT, and C_P? What parts of the calorimeter contribute to the total value of the constant C_P? Has any account been taken of the heat capacity of the thermometer?

4. Without referring to the text of the experiment, briefly describe the method of determining the calorimeter constant.

5. Without referring to the text of the experiment, briefly describe the graphical method of determining ΔT for the reaction. Why doesn't one simply take the temperature of the solutions just before and just after mixing to determine ΔT?

6. Why is it important that the acid and base solutions be at the same temperature before mixing?

7. Why is the normality of the base solution 2.05 while the normality of the acid solution is 2.00? Would it make any difference if the acid were 2.05 N and the base were 2.00 N?

8. The instructions of the experiment stress that all containers must be thoroughly drained of excess water between trials. This is not just for neatness. Why is this precaution important?

9. What kind of error in ΔT (i.e., over- or underestimation) would be most likely to result from inadequate stirring of the solutions after they are poured together? To answer this consider the likely appearance of the temperature versus time plot for a stirred versus unstirred reaction.

10. Suppose for a set of experiments ΔT was found to be $+15.4°C$, $+16.0°C$, $+15.9°C$, and the calorimeter constant is 1.4 cal deg^{-1}. Calculate the value of ΔH_{neut} and the standard deviation of this quantity.

EXPERIMENT

33

Determination of ΔH of Dissolution for $NH_4X(s)$

I. INTRODUCTION*

The present experiment is very similar to Experiment 32 except that ΔH will be measured for the process**

$$NH_4X(s) + H_2O(\text{excess}) \longrightarrow NH_4^+(aq) + X^-(aq)$$

where $NH_4X(s)$ is some salt of NH_4OH in the form of a solid. Unlike neutralization reactions, the ΔH of dissolution of salts (referred to as ΔH_{sol} in what follows) can be either positive or negative (endothermic or exothermic, respectively). Hence positive or negative values of ΔT can be observed. Also the magnitude of ΔH_{sol} is generally not so great as $\Delta H_{neutralization}$, so the ΔT values are not so large as for neutralization reactions. Another difficulty is that the salts do not dissolve immediately upon addition to water, and the salt-water mixture must be swirled reasonably vigorously for several minutes to assure complete dissolution. Consequently the determination of ΔH_{sol} is generally less accurate than $\Delta H_{neutralization}$. If laboratory time permits it is suggested that both Experiment 32 and Experiment 33 be done (each can be done in one three-hour laboratory period if all goes well).

In the following procedure, reference is made to Experiment 32 in which the basic techniques for the experiment are presented.

II. PROCEDURE

Determination of Calorimeter Constant

Follow the procedure in Experiment 32.

---

*The Introduction to Experiment 32 should be read before doing this experiment.

**Other cation salts may be substituted if desired.

Determination of ΔH_{sol} for NH$_4$X(s)

1. Weigh out approximately 5 g (to within ± 0.02 g) of the NH$_4$X(s) sample (unknown species).

2. Measure the temperature of 200 ml of distilled water for several minutes at ~30 sec intervals and record the values. For best results a precise alcohol thermometer ($\pm 0.02°C$) or the equivalent should be used because the temperature changes from dissolution will generally be rather small.

3. Add the 5 g sample of NH$_4$X(s) all at once and swirl reasonably vigorously. Record the temperature immediately after adding the NH$_4$X(s) and at ~30 sec intervals. Record this data. Note that ΔT may be negative for dissolution.

4. Construct a temperature versus time graph like Figure 32.1 and determine ΔT. For best results three determinations of ΔT should be carried out. Rinse all glassware and the styrofoam cup and drain carefully between determinations.

III. DATA ANALYSIS

Assuming a heat capacity of 0.96 cal g^{-1}deg^{-1} for the solution, and taking the total mass of the solution to be the sum of the mass of solute plus 200 g for the water, calculate ΔQ for each determination as follows:

$$\Delta Q = [(C_P)_{cal} + (C_P)_{sol}] \Delta T_{measured}$$

where $\quad (C_P)_{cal}$ = calorimeter constant (cal deg^{-1})

$\qquad (C_P)_{sol}$ = (0.96)(mass of solution)(cal deg^{-1}).

Then determine the enthalpy of dissolution per *gram** for each trial as follows:

$$\Delta H_{sol} = -\frac{\Delta Q}{(\text{mass of solute})}.$$

Obtain the average ΔH for all your trials and determine σ, the standard deviation (see Appendix 1). Report the individual values of ΔH and $\langle \Delta H \rangle \pm \sigma$.

IV. ERROR ANALYSIS

Calculate the maximum uncertainty in ΔH_{sol} that arises from the following sources:

1. temperature reading errors of $\pm 0.1°C$;

---

*It is assumed that the molecular weight of NH$_4$X is unknown, such that the ΔH_{sol} per gram must be calculated. If the identity of NH$_4$X is known, you may report ΔH_{sol} per mole instead.

2. errors in the weight of the solute of ± 0.1 g;

3. inaccuracies in the stated heat capacity of the solution of ± 0.02 cal g^{-1} deg^{-1};

4. inaccuracies in the volume of water used for the dissolution of ± 10 ml;

5. a ± 20 percent error in the determined value of $(C_P)_{cal}$ (i.e., multiply $(C_P)_{cal}$ by 1 ± 0.2);

6. the error if 0.4 g of solute does not dissolve during the period for which the temperature is measured.

Based upon the above, or other errors not considered, estimate a reasonable value for the uncertainty in ΔH_{sol}, and compare with σ. If there is a serious discrepancy between these values, try to analyze the reason(s).

NAME   DATE

SECTION

SELF-STUDY QUESTIONS

1. What is a calorimeter?

2. If ΔT for a reaction is negative is the reaction endothermic or exothermic?

3. What is the mathematical relationship between ΔQ, ΔT, and C_P? What parts of the calorimeter contribute to the total value of the constant C_P? Has any account been taken of the heat capacity of the thermometer?

4. Without referring to the text of the experiment, briefly describe the method of determining the calorimeter constant.

5. Without referring to the text of the experiment, briefly describe the graphical method of determining ΔT for the reaction. Why doesn't one simply take the temperature of the solutions just before and just after mixing to determine ΔT?

6. Why is it important to swirl the solute-solvent mixture during the course of the temperature measurement?

7. In this experiment, is it important that the temperature of the water before addition of the solute be the same within $\pm 1°C$ for each trial, or can it differ by more than this amount? Explain your answer.

EXPERIMENT

34

Semi-quantitative
Determination of the
Temperature
Dependence of the
Solubility of a Lead
Compound

I. INTRODUCTION

One of the commonly encountered examples of an equilibrium process is the dissolution of a partially soluble salt. For example, consider a lead compound of the general formula PbX_2 where dissolution involves the process

$$PbX_2 \text{ (solid)} \xrightarrow{\text{water}} Pb^{+2} \text{ (aq)} + 2X^- \text{ (aq)}.$$

For a given temperature the equilibrium between solvated ions and solid PbX_2 obeys the relation*

$$[Pb^{+2}][X^-]^2 = K_{sp}(T)$$

so long as there is solid PbX_2 in contact with the solution. The solubility product has been written $K_{sp}(T)$ to emphasize the fact that this equilibrium constant is temperature dependent just like all equilibrium constants. It is the purpose of this experiment to evaluate $K_{sp}(T)$ at several different temperatures. Using a standard thermodynamic relationship, $d \log K_{sp}(T)/d(1/T) = -(\Delta H/2.3R)$, the ΔH for dissolution of PbX_2 (solid) can be estimated.

In this experiment a saturated solution of $PbX_2(s)$ will be established at various temperatures (using a water heat bath) between room temperature and approximately 80°C. After equilibrium is established, 10 ml of this solution will be withdrawn and placed in a clean test tube. After 3–5 samples have been obtained they will be redissolved in a boiling water bath and the Pb^{+2} will be precipitated by addition of H_2SO_4. The $PbSO_4$ will be dried and weighed and the number of moles of Pb^{+2} computed. From this information the value of K_{sp} at the various

---

*More correctly the concentrations of Pb^{+2} and X^- should be replaced by the activities of Pb^{+2} and X. See the Appendix to this experiment.

temperatures will be determined. For example, if at a particular temperature X $mM(mM =$ millimoles, moles $\times 10^{-3})$ of $PbSO_4$ are recovered then the concentration of Pb^{+2} in contact with PbX_2 (solid) was $X/10\,mM$/ml ($=$ moles/liter) and since $[X^-] = 2\,[Pb^{+2}]$

$$K_{sp} = 4\left(\frac{X}{10}\right)^3$$

The methods to be employed are not as accurate as normal quantitative analysis and hence the determination is stated to be semi-quantitative. The primary approximations used are: (1) a 10 ml pipette will be used at temperatures from room temperature to 80°C but it is usually calibrated at 20°C only; (2) the precipitate will be recovered by decanting and washing rather than filtration, which could result in unwanted absorbed material on the surface being retained; (3) a gravimetric method is used even though the weights of precipitate will be small, from 25 to 250 mg.

There are several places in the following procedure where serious errors can be made. Special attention should be paid to the following:

1. Allow adequate time for the saturated solution to come to equilibrium. Stir the solution frequently. Use a stirring rod or a long dropper to continually mix the solid PbX_2 with all portions of the solution (e.g., pick up crystals of PbX_2 from the bottom of the test tube with the dropper and allow them to drop through the top portion of the solution; without taking this precaution a significant change in concentration can be established in the test tube.

2. When withdrawing the 10 ml sample use care to avoid picking up crystals of PbX_2. Allow plenty of room between the end of the pipette and the solid. Make sure all crystals have settled to the bottom of the tube if the mixture has recently been stirred.

3. When decanting the liquid from above the $PbSO_4$ solid (after precipitation with H_2SO_4), be careful not to lose any solid $PbSO_4$. Leaving a little extra water or ethanol will not hurt the results since they are evaporated off at the end of the last step.

II. PROCEDURE

1. Clean, dry, and label 4 or 5 150 $\times$ 15 mm test tubes.*
2. Add approximately 1–1½ g of PbX_2 sample to a 150 $\times$ 25 mm test tube with approximately 25 ml of distilled water. Place a thermometer and long dropper

---

*Label test tubes with a pencil mark or other identification that will be stable in a drying oven (110°C) and in the presence of water. If possible, the weight of the clean and *dry* test tubes should be recorded at this time. This can be done later if it is more convenient. It may be preferable to mark the 150 $\times$ 15 mm test tube at the level occupied by 10 ml of water, and suspend this test tube in the water bath. Then the saturated PbX_2 solution can be transferred to this test tube using the long dropper. This procedure avoids the transfer of the hot solution using a 10 ml pipette (which is awkward) but is less accurate.

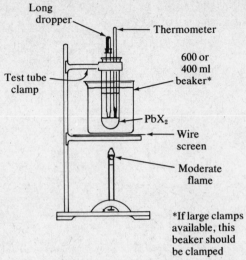

Fig. 34.1 *Water bath with saturated PbCl₂ solution.*

in the test tube. Support the test tube in a water bath, as illustrated in Figure 34.1, with a good support for the water bath. Begin heating with a moderate flame with frequent agitation and mixing of the test tube contents with the dropper.

3. As the temperature rises take samples of the solution with a 10 ml pipette and place in a 150 × 15 mm test tube. Note the temperature of the sample. Before withdrawing the sample rinse the pipette with water from the water bath several times to warm it.

NOTE: For all operations with the pipette use a pipette bulb as illustrated in Figure 34.2 to avoid accidents.

Before withdrawing a sample, be sure the PbX₂ solid has completely settled and that *no crystals are drawn along with the liquid.* After each sample is withdrawn add more distilled water and continue heating. Allow at least 10–15 minutes between sampling. Pick temperatures between 35° and 80°C and obtain 3 to 4 samples.

4. After all the samples have been taken, the large test tube may be removed from the water bath and the sample tubes placed in the hot (90°–100°C) water. Stir each tube until all solids are dissolved (be very careful not to cross-contaminate; wipe the stirring rod after each use). After all solids are dissolved add ~2 ml of 6N H₂SO₄. A heavy white precipitate (PbSO₄) will form immediately. Remove the test tubes from the water bath and allow to stand overnight or longer (covering is not required but avoid any contamination).

NOTE: If the contents of the large test tube have returned to room temperature for at least 15 minutes, another 10 ml sample can be taken, or the room-temperature sample can be taken and precipitated at the next laboratory period.

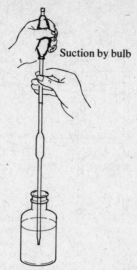

Suction by bulb

Fig. 34.2 *Using a suction bulb with a pipette.*

5. At the next laboratory period the supernatent liquid over the $PbSO_4$ can be decanted for each sample (a long dropper can be used to advantage here). Wash the precipitate with 95 percent ethanol (5–7 ml), and allow the precipitate to settle. Decant and place the test tubes in a drying oven for an hour or more (it is often advisable in larger laboratory sections to allow the ethanol to evaporate into the air between periods and dry in an oven during the next period). After the test tubes are dried and have cooled they may be weighed. If the test tubes have been previously weighed the weight of $PbSO_4$ in each sample may be calculated. If the test tubes have not been weighed, they should be carefully cleaned (use a test tube brush), oven dried, and weighed after dry and cool. Be careful that your identification marks on the test tubes are not wiped off during these processes.

6. The mass of $PbSO_4$ obtained for each sample should be recorded in a table along with the temperature at which the sample was taken. The number of millimoles of Pb^{+2} in the original sample is given by

$$\frac{\text{number of millimoles}}{Pb^{+2}\ \text{dissolved}} = X = \frac{\text{weight of } PbSO_4\ \text{(mg)}}{303.2\ \text{mg/mmole}}.$$

Hence the concentration of Pb^{+2} is given by X mmoles/10 ml and

$$K_{sp} = 4\left(\frac{X}{10}\right)^3$$

(see section IIa). Tabulate K_{sp}, log K_{sp}, and $1/T$ (in degrees absolute) for each sample.

NOTE: In general, K_{sp} should increase with T.

III. DATA ANALYSIS

Calculation of $\Delta H°$ Solubility

In general for any equilibrium constant K_{eq} one may write

$$\log K_{eq} = -\frac{\Delta G°}{2.3RT} = -\frac{\Delta H°}{2.3RT} + \frac{\Delta S°}{2.3R}$$

If $\Delta H°$, $\Delta S°$ vary slowly with temperature, a plot of $\log K_{eq}$ versus $1/T$ (T expressed in degrees Kelvin) will yield a straight line* with slope $-\Delta H°/2.3R$. Taking the K_{sp} to be an example of an equilibrium constant, plot $\log K_{sp}$ (to base 10) versus $1/T$ (see graph paper provided) and determine $\Delta H°$ for the process $PbX_2(s) \longrightarrow Pb^{+2}(aq) + 2X^-(aq)$ by obtaining the slope of $\log K_{sp}$ versus $1/T$ using a least-squares fit to your data.

The result of this calculation is approximate because the interaction between solvated ions causes deviations from a simple equilibrium expression $K_{sp} = [Pb^{+2}][X^-]^2$. For example, the solubility of $PbX_2(s)$ will change if other salts (such as $NaNO_3$) are dissolved in the solvent even though the other ions do not directly enter into the $PbX_2(s) \rightleftharpoons Pb^{+2}(aq) + 2X^-(aq)$ equilibrium. The correction for this effect is complicated and will be described in Appendix 33 for those students carrying out a more detailed analysis of these data.

IV. ERROR ANALYSIS

Calculate the maximum uncertainty in K_{sp} at several temperatures from the following errors. Enter the maximum estimated uncertainty in the K_{sp} as error bars on your plot of $\log K_{sp}$ versus $1/T$:

1. a weighing error of ± 10 mg on the weight of $PbSO_4$ precipitate;

---

*Use least-squares fit, Appendix 2.

2. the use of an insufficient amount of 6 N H_2SO_4, such that the precipitate weighed was actually 50 percent $PbSO_4$ and 50 percent $PbCl_2$;

3. volumetric errors of ±0.5 ml in taking the 10 ml sample of saturated $PbCl_2$ solution (you will have to use your value of K_{sp} to estimate the effect of this error);

4. accidentally including a 10 mg crystal of $PbCl_2$ in the 10 ml sample of saturated $PbCl_2$ solution;

5. errors in temperature on the order of $\pm2°C$ (you will have to use your results of the temperature dependence of K_{sp} to estimate the effect of this error).

List the types of experimental errors of technique that could lead to the types of errors considered above. In addition, discuss the effect on K_{sp} values of inadequate stirring of the saturated solution, taking solution samples before the solution has equilibrated, using a 10 ml pipette that is calibrated at 20° or 25°C for a solution at 50°C or higher. Are there other sources of error that you can analyze?

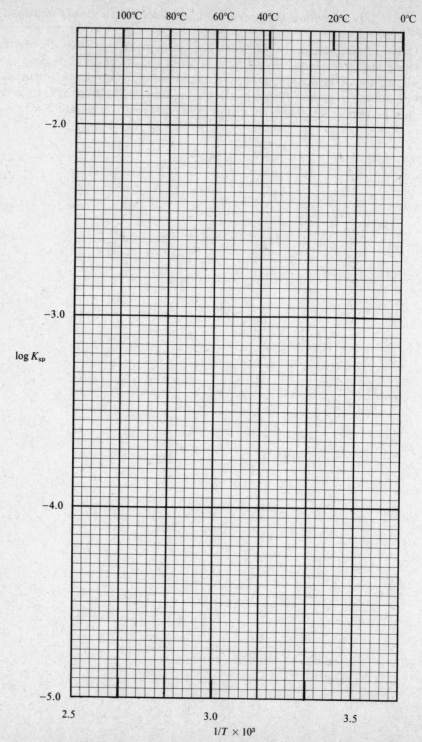

Fig. 34.3 *Graph for data, least-squares line, for Experiment 34.*

NAME _____ **DATE** _____

SECTION _____

SELF-STUDY QUESTIONS

1. What is the form of the solubility product for $PbCl_2$ and $PbSO_4$?

2. If the solubility of a solid increases with temperature, is $\Delta H_{solution}$ for that solid positive (endothermic) or negative (exothermic)?

3. In this experiment, why is Pb^{+2} precipitated as $PbSO_4$ rather than just weighing the solid $PbCl_2$ directly?

4. Why is it necessary to wait 10–15 minutes after each new temperature of the water bath is obtained? What is the point of continually stirring the crystals and solution?

5. Why is the $PbSO_4$ solid washed with 95 percent ethanol before drying in the oven? Is this step absolutely necessary?

6. Why do you think the $PbCl_2(s)$ is redissolved by heating before precipitation with 6 N H_2SO_4 is carried out? What error might occur if H_2SO_4 were simply added to the room temperature $PbCl_2$ solution with $PbCl_2$ crystals remaining undissolved?

7. Suppose 185.4 mg of $PbSO_4$ is obtained from one 10 ml sample of $PbCl_2$ solution. What is the solubility product of $PbCl_2$ implied by this value?

APPENDIX 34. THE ACTIVITY COEFFICIENT CORRECTION TO K_{sp}

Expressing the dissolution of $PbX_2(s)$ by the reaction

$$PbX_2(s) = Pb^{+2}(aq) + 2X^-(aq)$$

and $K_{sp} = [Pb^{+2}][X^-]^2$ is an oversimplification that ignores the rather strong interaction between the solvated ions of opposite charge in solution. These interactions are more important as the concentration of the electrolytic solution increases but are still appreciable at the rather low concentrations of electrolyte encountered in this experiment. The thermodynamically correct expression for the solubility product is of the form

$$K_{sp}^{thermo} = A_{Pb^{+2}}(A_{X^-})^2$$

where A is the *activity* of the ionic species in solution. The activity of an ion approaches the concentration of that ion as the concentration becomes very low such that it is the usual convention to write

$$A_{Pb^{+2}} = \gamma_{Pb^{+2}}[Pb^{+2}]$$

where $\gamma_{Pb^{+2}}$ is the *activity coefficient* of Pb^{+2} and is a complicated function of concentration with the properties

$$\gamma_{Pb^{+2}} \longrightarrow 1.0 \text{ as } [Pb^{+2}] \longrightarrow 0$$
$$0 \leq \gamma_{Pb^{+2}} \leq 1.0 \text{ for all concentrations.}$$

The activity of $X^-(aq)$ is expressed in terms of γ_{X^-} in the same way.

At first glance the replacement of K_{sp} by K_{sp}^{thermo} seems to be a needless complication. It is found experimentally, however, that the solubility of any slightly soluble salt depends not on the concentration of the ions in the salt alone, but rather on the concentration of all electrolytes in the solution. In the example being considered here the solubility product of $PbX_2(s)$ would be significantly increased by addition of a salt like $NaNO_3$, even though Na^+ and NO_3^- ions play no direct role in the $PbX_2(s) = Pb^{+2}(aq) + 2X^-(aq)$ equilibrium. This is properly expressed in K_{sp}^{thermo} because addition of $NaNO_3$ would affect the *activities* of $Pb^{+2}(aq)$ and $X^-(aq)$, but the product of the activities would still equal the constant K_{sp}^{thermo}.

In general obtaining the value of γ is rather difficult. For the present case a theory of electrolytic solutions, presented by Debye and Hückel in 1923, allows a reasonably accurate estimate of γ to be made. The derivation of the expression that follows is quite complex, arising as it does from a combination of electrostatics and thermodynamics. The final result for water is

$$\log \gamma = -\frac{0.509}{2.3}|Z_+Z_-|\left(\frac{298}{T}\right)^{3/2} I^{1/2}$$

where γ = average activity coefficient for either positive or negative ion
  Z_+, Z_- = charge on each type of ion, e.g., $Z_+ = 2$ for Pb^{+2}, $Z_- = 1$ for X^-
  I = ionic strength

$$= (1/2) \sum_{i=1}^{N} c_i Z_i^2$$

c_i = concentration of ith ionic species (M)
Z_i = charge of ith species

Note that

$$I = \frac{1}{2}[X(2^2) + (2X)(1)] = 3X$$

since $c_{Pb^{+2}} = X, c_{X^-} = 2X$.

Using the activity coefficients, we find that

$$K_{sp}^{thermo} = [Pb^{+2}]\gamma_{Pb^{+2}}[X^-]^2\gamma_{X^-}^2 = K_{sp}\gamma^3$$

so

$$\log K_{sp}^{thermo} = \log K_{sp} - 3\log\gamma$$

For the present case, the Debye-Hückel expression may be used for γ. With this correction $\log K_{sp}^{thermo}$ versus $1/T$ can be computed. The ΔH_{sol}° can be obtained from the slope of this plot. Compare the ΔH_{sol}° result with and without the activity coefficient correction. Don't be surprised if the ΔH_{sol}° changes very little because the heat of solution (ΔH_{sol}°) is much less dependent on the concentration of other ions than the entropy of solution (ΔS_{sol}°).

SECTION J

Electrochemistry

EXPERIMENT

35

Redox Processes and Faraday's Law

I. INTRODUCTION

Redox processes are chemical reactions in which substances undergo a change in oxidation number. For example, the dissolution of metallic zinc in aqueous hydrochloric acid (Eq. 1)

$$Zn + 2HCl \longrightarrow ZnCl_2 + H_2 \tag{1}$$

involves oxidation and reduction because zinc increases in oxidation number (from zero in the atom to 2+ in $ZnCl_2$) and hydrogen decreases in oxidation number (from 1+ in HCl to zero in H_2). Remember that all redox reactions can be considered to occur because of electron transfer. One process (oxidation) involves a loss of electrons whereas the other (reduction) involves a gain of electrons. Thus, Equation 1 can be broken down into two steps (half-reactions), one (Eq. 2) supplying the electrons required for the other (Eq. 3).

$$Zn \longrightarrow Zn^{2+} + 2e^- \tag{2}$$

$$2H^+ + 2e \longrightarrow H_2 \tag{3}$$

One important point should be noted in redox processes, viz., the total number of electrons produced in one half-reaction is the number of electrons consumed in the other half-reaction; there is no excess charge left over when the process is completed. Since equations such as Equation 2 and Equation 3 are balanced with respect to mass and charge, they provide the basis for establishing the equivalency between mass and charge. Thus, Equation 2 indicates that one mole of zinc produces two moles of electrons when it dissolves. From a chemical point of view, a mole of electrons is a fundamental amount of electricity in the same way that a mole of a substance is a fundamental amount of matter; both represent the same number of particles, i.e., Avogadro's number. A mole of electrons is called a faraday, after Michael Faraday, who discovered the basic laws which relate electricity to matter. One faraday of electricity will produce or consume an equivalent (in terms of electron transfer) amount of matter; this amount is called the equivalent weight of the substance. For example, from Equation 2 we see that one mole of electrons (which is Avogadro's number and hence, also called one faraday) is produced by ½ mole (65.37/2 = 32.69 g) of zinc. Equation 3 indicates that each mole of electrons produces ½ mole of H_2 (2.00/2 = 1.00 g). Thus, when zinc and hydrogen are involved in a redox process 32.69 g of zinc and 1.00 g of H_2 are equivalent quantities in the sense that they are both involved in 1 mole of electrons. 32.69 g and 1.00 g are said to be the *equivalent weights* of zinc and hydrogen, respectively.

Reactions 2 and 3 occur when zinc and aqueous HCl are brought together, the electron transfer occurring at the metal surface. It is also possible to arrange reactions 2 and 3 to occur in an *electrochemical* system where the electrons travel through an external circuit.

In this experiment we will reduce hydrogen ions (Eq. 3) in one electrochemical cell at the same time that we oxidize an unknown metal in another electrochemical cell. We shall arrange to have the cells in series so that the same current (same number of electrons) passes through both cells. By measuring the amount of hydrogen gas liberated, we have a measure of the number of electrons passed through the cell. This information together with the weight loss of the metal electrode (Eq. 4)

$$M \longrightarrow M^{n+} + ne^- \tag{4}$$

should provide sufficient information to identify the metal.

II. PROCEDURE

Set up the apparatus as shown in Figure 35.1. A burette, which will be used to measure the volume of gas liberated, is placed in a beaker containing 100 ml of $1M$ H_2SO_4. The other 100 ml beaker contains 100 ml of $0.5M$ KNO_3 solutions. A piece of heavy nichrome wire acts as an electrical connector between the beakers. Nichrome is used because this alloy is relatively inert and the components will not become involved in the redox processes which will occur upon electrolysis. The electroactive electrodes are a piece of heavy copper wire coated with waterproof insulation in the H_2SO_4 solution and a piece of the unknown metal in the KNO_3 so-

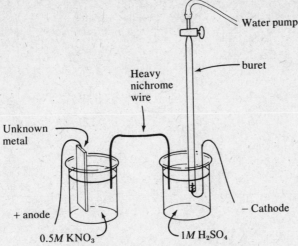

Fig. 35.1   *Apparatus for Experiment 35.*

lution. The current will be supplied by a suitable power source provided by your instructor.

Use the following procedure to conduct the electrolysis.

1. Draw some H_2SO_4 solution to the buret tip using a piece of rubber tubing and a water pump. Make certain the buret stopcock is greased before you draw the acid up. The acid level in the buret should be near the stopcock but on the graduations. Check the level after a few minutes to make certain that the stopcock does not leak. Record the acid level in the buret.

2. Make certain the copper wire cathode is well within the buret. There should be no bare wire exposed to the solution outside of the buret.

3. Weigh, and record the weight of the unknown metal.

4. Attach the metal to the anode using a spring clip.

5. Start the electrolysis by immersing the anode in the KNO_3 solution. Do not immerse the spring clip in the solution. Hydrogen gas should immediately be generated at the cathode and collected in the buret. Collect about 50 ml of gas and record the liquid level in the buret. Disconnect the power source, record the temperature of the room and the atmospheric pressure. You may see a cloudyness appearing at the anode while the electrolysis proceeds. This will not affect your experiment.

6. Open the stopcock on the buret, draw acid to the top of the graduations, read the buret, reconnect the power source, and generate approximately 50 ml of hydrogen again. Record the level of acid in the buret, the pressure and temperature of the room.

7. Remove the metal anode, rinse it with distilled water, and quickly with 1*M* acetic acid, again with distilled water, and then with acetone. Let the acetone evaporate in air and weigh the anode; record the weight. If the anode has acquired a flaky coating, you will have to scrape this off, and then rinse and dry the electrode.

Table 35.1
The Vapor Pressure of Water
Over a 1 M H_2SO_4 Solution

T, °C	P, torr
0	4.38
5	6.30
10	8.80
15	12.3
20	16.6
25	22.4
30	30.0
35	40.1
40	52.9
45	68.1
50	88.5

III. DATA ANALYSIS

Assume that the following information was obtained in this experiment. Upon electrolysis the anode lost 0.232 g while 27.2 ml $H_2(g)$ (measured at 25°C and 752 torr) are produced.

1. Calculate the number of moles of hydrogen formed using the ideal gas law.

$$PV = nRT$$

Since the hydrogen was collected over $1M$ H_2SO_4 it contains water vapor; we must correct for the partial pressure of water over $1M$ H_2SO_4 (see Table 35.1). At 25°C the vapor pressure of water over $1M$ H_2SO is 22.4 torr. Thus, the pressure of hydrogen is $752 - 22.4 = 730$ torr. Rearranging the ideal gas law for the number of moles of gas gives

$$n = \frac{PV}{RT}$$

Substituting the appropriate quantities yields

$$n = \frac{(730 \text{ torr})(0.0272 \; l)}{\left(62.36 \dfrac{\text{torr } l}{\text{mol °K}}\right)(298°K)} = 1.07 \times 10^{-3} \text{ moles}$$

2. Calculate the number of faradays of electrons passed through the circuit. Since 1 mole of H_2 requires 2 faradays, the number of faradays passed is

$$2 \times 1.07 \times 10^{-3} = 2.14 \times 10^{-3}$$

But this must be the number of equivalents of metal oxidized (lost) at the anode. Thus the equivalent weight of the metal is given by

$$\text{eq. wt.} = \frac{\text{g. metal oxidized}}{\text{faradays passed}} = \frac{0.232 \text{ g}}{2.14 \times 10^{-3}} = 108 \text{ g}$$

The metal must have an atomic weight which is a whole number multiple of 108. Inspection of the atomic weights of the elements indicates that the metal is probably silver.

In a similar fashion, calculate the equivalent weight of your unknown metal.

IV. ERROR ANALYSIS

Estimate the maximum uncertainty in the equivalent weight of your metal that arises from the following sources.

1. An error of $\pm 3°$ is made in the temperature at which H_2 is collected.

2. The volume of H_2 is determined at 25°C but a correction for the vapor pressure of water is not made.

3. The barometric pressure is in error by ± 7 torr.

4. The weight of the anode is in error by ± 0.001 g.

5. The concentration of the KNO_3 is in error by $\pm 0.1\,M$.

6. The concentration of H_2SO_4 is in error by $\pm 0.1\,M$.

Describe the error expected if the following experimental conditions obtained.

7. The copper anode was not insulated.

8. The spring clip attached to the anode was immersed while the electrolysis occurred.

NAME _____   **DATE** _____

SECTION _____

SELF-STUDY QUESTIONS

1. Give a balanced half-reaction for the anodic oxidation of magnesium to its most stable oxidation state.

2. How many faradays of electricity are required to oxidize a mole of magnesium?

3. If magnesium is oxidized while hydrogen ion is reduced and 0.05 faradays pass through the system:
   a. How many electrons are involved?
   b. What weight of magnesium is oxidized?
   c. What volume of hydrogen (measured at STP) is produced?

4. Describe a non-electrochemical method of determining the equivalent weight of magnesium.

EXPERIMENT

36

Voltaic Cells and the Nernst Equation: Determination of $\Delta G°$, K_f for $Cu(NH_3)_4^{2+}$

I. INTRODUCTION

One of the fundamental reactions in chemistry is a redox reaction in which an electron transfer occurs, for example:

$$Zn(s) + Cu^{2+}(aq) = Zn^{2+}(aq) + Cu(s). \tag{1}$$

In this example, Zn transfers two electrons to the Cu^{2+} ion.* This reaction can be observed by placing a strip of Zn metal in a solution of Cu^{2+} ($CuSO_4$ solution, for example), in which case a reddish deposit of Cu will quickly plate out on the zinc. The above reaction can also be made to drive an electron current through an external circuit. Such a device is known as a *voltaic cell* (also galvanic cell).** The schematic diagram for a cell using reaction 1 (with $ZnSO_4$, $CuSO_4$ electrolytes) is illustrated in Figure 36.1.

An extremely important feature of the voltaic cell is the porous partition which prevents gross mixing of the $CuSO_4$ solution with the $ZnSO_4$. If these solutions were mixed, reaction 1 would occur without any current having to pass through the external circuit. On the other hand, the partition must be able to pass ions at a finite rate because the Zn^{2+} ions produced in the left compartment must be balanced by incoming SO_4^{2-} ions. Similarly, Cu^{2+} ions in the right compartment are being lost as the cell operates and SO_4^{2-} ions must leave to prevent an overall charge imbalance.*** A variety of methods are used to produce this porous partition including glass frits (such as used for filtering crucibles), paper or cellulose

---

*The direction in which a reaction such as 1 goes depends on the position of the two elements in the standard electrode potentials (see Table 36.1). In this case the reaction goes from left to right.

**After Alessandro Volta (1800) and Luigi Galvani (1780), respectively.

***Each compartment maintains electroneutrality (equal positive and negative charge) to a very high degree. If there were no ionic exchange between compartments (i.e., a non-porous partition), only an immeasurably small current would flow before the charge imbalance prevented further electron transfer.

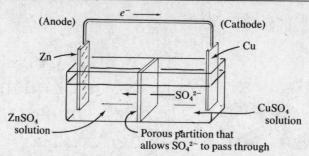

Fig. 36.1 *Schematic diagram for a cell using reaction with $ZnSO_4$ and $CuSO_4$ electrolytes.*

membranes, or the *salt bridge* (a solid support containing a solution of electrolyte (e.g., NaCl or KCl) that allows ions to move from one compartment to the other). A common method of constructing a salt bridge is to combine agar (a gelatine-like substance) with the electrolyte. In this experiment we use a piece of filter paper soaked in electrolyte as a bridge (see Fig. 36.2 for the type of cell used here).

There are two properties of a voltaic cell that are generally of interest, the *current* (amps = coulombs/sec)* and the *voltage* (volts).** The former quantity depends on a number of properties of the cell that are not of interest to the present experiment, such as the size of the electrodes, the efficiency of the salt bridge, the internal resistance of the cell, and the voltage developed between the two electrodes. The voltage depends on the chemical identity of the electrode materials and the concentration and type of electrolyte in each compartment. Ignoring for the moment the effect of the electrolyte, the voltage developed between a pair of metal electrodes can be predicted by referring to a table of standard electrode potentials, such as Table 36.1. These potentials are listed for hypothetical half-reactions and have meaning only when two half-reactions are combined for a redox reaction, i.e.,

$$
\begin{aligned}
Cu^{2+} + 2e^- &= Cu(s) & E^\circ &= +0.337 \\
Zn(s) &= Zn^{2+} + 2e^- & E^\circ &= +0.763 \\
\hline
& & E^\circ_{cell} &= +1.100 \text{ V}
\end{aligned}
\tag{2}
$$

Note that the zinc reaction is written as zinc metal oxidation, while the standard potentials in Table 36.1 are given as metal ion reductions. *This reversal of the reaction requires that the sign of E° be reversed.* The quantity E°_{cell} is known as the standard emf of the cell.

The concentration of the electrolyte has a significant effect on the voltage of the cell. A discussion of this relation is given in Appendix 36. For the moment, we will simply state that the standard potential for the half-reaction should be written

$$
E = E^\circ - (2.3)\frac{RT}{nF} \log \frac{a_{M(s)}}{a_{M^{n+}}}
\tag{3}
$$

---

*One mole of electrons is the unit known as the faraday, approximately 96,500 coulombs/mol.

**For practical applications the *power* of the battery is of greatest importance, which is current × voltage = joules.

Table 36.1
Standard Electrode Potentials at 25°C

Half-reaction	$°(volts)$
$Li^+ + e^- = Li$	-3.045
$K^+ + e^- = K$	-2.925
$Ba^{2+} + 2e^- = Ba$	-2.906
$Ca^{2+} + 2e^- = Ca$	-2.866
$Na^+ + e^- = Na$	-2.714
$Mg^{2+} + 2e^- = Mg$	-2.363
$Zn^{2+} + 2e^- = Zn$	-0.7628
$Fe^{2+} + 2e^- = Fe$	-0.4402
$Pb^{2+} + 2e^- = Pb$	-0.126
$2H^+ + 2e^- = H_2$	0.0
$Cu^{2+} + 2e^- = Cu$	$+0.337$
$Cu^+ + e^- = Cu$	$+0.521$
$Fe^{3+} + e^- = Fe^{2+}$	$+0.771$
$Ag^+ + e^- = Ag$	$+0.7991$
$Br_2 + 2e^- = 2Br^-$	$+1.0652$
$O_2 + 4H^+ + 4e^- = 2H_2O$	$+1.229$
$Cl_2 + 2e^- = 2Cl^-$	$+1.3595$
$F_2 + 2e^- = 2F^-$	$+2.87$

where

E = observed emf of half-reaction

$a_{M(s)}$, $a_M{}^{n+}$ = activities of M(s), M^{n+}(aq)

F = Faraday's constant, $\dfrac{23{,}060 \text{ cal}}{\text{V-mol*}}$

n = number of electrons transferred

R = universal gas constant, $\dfrac{1.987 \text{ cal}}{\text{deg-mol}}$

T = temperature in degrees Kelvin

Note that $E = E°$ only when all activities are unity. For a pure metal, $a_{M(s)} = 1$, so

$$E = E° + \frac{RT}{nF} \, 2.3 \log a_{M^{n+}} \tag{4}$$

$$E = E° + \frac{0.05916}{n} \log a_{M^{n+}}.$$

Thus for the Zn/Cu redox reaction, combining the two half-reactions yields

$$E_{cell} = E°_{Cu} - E°_{Zn} + \frac{0.05916}{2} \log (a_{Cu^{2+}}) - \frac{0.05916}{2} \log (a_{Zn^{2+}}) \tag{5}$$

$$= E°_{cell} - \frac{0.05916}{2} \log \frac{a_{Zn^{2+}}}{a_{Cu^{2+}}}$$

---

*Faraday's constant was stated earlier as approximately 96,500 coulombs/mol (actual value 96,487 coulombs/mol). The conversion between these units is as follows: coulombs × volts = joules, so coulombs/mol = joules/volt-mol. The conversion between joules and calories is 0.2390057 cal/joule, so F = 96,487 coulombs/mol = 96,487 joules/volt-mol × 0.2390057 cal/joule = 23,061 cal/volt-mol.

where the second form is the *Nernst equation* at 25°C. The most general form of the Nernst equation is

$$E_{\text{cell}} = E^{\circ}_{\text{cell}} - \frac{0.05916}{n} \log \left[\frac{(a_Y)^y (a_Z)^z}{(a_A)^a (a_B)^b} \right] \tag{6}$$

for the cell reaction

$$a\text{A} + b\text{B} = y\text{Y} + z\text{Z}. \tag{7}$$

There are several conventions for representing a cell that must be adhered to if confusion is to be avoided. By convention, the *left electrode* will be the electrode where oxidation occurs and is referred to as the *anode*. For the zinc electrode in Figure 36.1 we write

$$\text{Zn(s)} = \text{Zn}^{2+} + 2e^- \qquad (\text{Zn(s) is oxidized}).$$

For the *cathode* (the right electrode) we write

$$\text{Cu}^{2+} + 2e^- = \text{Cu(s)} \qquad (\text{Cu}^{2+} \text{ is reduced}).$$

The standard cell emf (by this convention) is

$$E^{\circ}_{\text{cell}} = E^{\circ}_{\text{right}} - E^{\circ}_{\text{left}}$$

and the cell is diagrammed as

$$\text{Zn}|\text{Zn}^{2+}(\text{conc.})\|\text{Cu}^{2+}(\text{conc.})|\text{Cu}$$

where each vertical line represents a phase boundary (i.e., electrode/solution for this case) and the double vertical line represents the salt bridge, which we presume to have no effect on the E_{cell} value.* The concentration of the electrolytes should be given for a complete description of the cell (normally the anion need not be specified unless there is complex ion formation, such as CuCl_4^{2-}). The choice of the electrode for the anode, according to this convention, is the material with more negative E° value in Table 36.1. By following this convention the calculated E°_{cell} will always be positive, and electrons will flow from the left electrode to the right. This is perhaps the most important aspect of the present set of conventions: if $E_{\text{cell}} > 0$ then electrons flow from left to right. If a "wrong" choice in designating the anode and cathode is made, such that $E_{\text{cell}} < 0$, then the flow of electrons is from right to left. *It is most important to grasp the direction of electron flow in a given cell in order to appreciate the nature of the chemical reaction that is taking place.*

There are several additional comments that should be made concerning cell emf's and the Nernst equation:

1. The effect of concentration on activity. In Equations 5 and 6 it is the activities of the ions or species that are to be used. In the limit of dilute solutions the activity of a species approaches its concentration, but for the present experi-

---

*The salt bridge will have some effect, but usually a relatively minor one, on the order of 0.050 V or less, *if* the bridge is properly constructed.

ments (concentration $\gtrsim 0.5\ M$) the deviation of activity from concentration is significant and difficult to compute.

2. The effect of current on E_{cell}. The computed E_{cell} values discussed above are essentially thermodynamic quantities and are appropriate only for ideal, chemically reversible conditions. This means that E_{cell} actually applies to the case when essentially no current is being drawn from the cell (the "open circuit" voltage). This requires that the voltage measuring device draw very little current ($\lesssim 10^{-3}$ amp), which in turn requires a potentiometer, a vacuum tube voltmeter, or the equivalent. If too much current is drawn from the cell, the measured voltage will be less than the calculated E_{cell}.

II. PROCEDURE

The determination of E°_{cell} for a cell in which a complex ion is formed will be given in the following section. In Experiment 37 the solubility product of a slightly soluble salt will be determined from the charge of voltage in a cell. Both experiments require the preparation of a cell following the instructions of section A.

A. Preparation of a Simple Voltaic Cell

Refer to Figure 36.2 for a sketch of the type of cell to be used. It is recommended that the two beakers be taped together or held together by a rubber band to minimize the possibility of spilling. All metal electrodes must be cleaned of oxide or other coatings by a light sanding with sandpaper or the equivalent. It is also quite important to remove any coating where the alligator clips make contact with the electrode. The electrolyte solutions are made to the concentrations specified by the experiment. The salt bridge is made by soaking a piece of filter paper

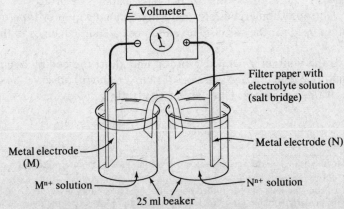

Fig. 36.2. *Experimental arrangement for voltaic cell.*

($\sim\frac{3}{8}$ in. $\times$ 3 in.) in an electrolyte solution (e.g., saturated NaCl or KCl). Instructions on proper use of the voltmeter will be given by your laboratory supervisor.*

B. Determination of E°_{cell} for $Cu|Cu(NH_3)_4^{2+}||Cu^{2+}|Cu$

In this experiment a cell will be constructed of the following type:**

$$Cu|Cu(NH_3)_4^{2+}(0.5M),NH_3(5.4M)||Cu^{2+}(0.5M)|Cu.$$

In this case the cell consists of two copper strips for electrodes, a 0.5 $CuSO_4$ solution in the right compartment, and a 1:1 mixture of 1.0 M $CuSO_4$ and concentrated NH_3 (14.8 M). The resulting solution is approximately 0.5 M in $Cu(NH_3)_4^{2+}$ and 5.4 M in free NH_3 (assuming that the formation of the complex ion is essentially quantitative).

1. Prepare or obtain the following solutions: 25 ml of 1.0 M $CuSO_4$, 5–10 ml of saturated NaCl or KCl solution, 10 ml of concentrated NH_3.

2. In the left (anode) cell combine 10 ml of the 1.0 M $CuSO_4$ solution with 10 ml of concentrated NH_3. After brief stirring there should be no light colored solid ($Cu(OH)_2$), but only the dark blue-violet $Cu(NH_3)_4^{2+}$ complex should exist.

3. In the right (cathode) cell combine 10 ml of 1.0 M $CuSO_4$ with 10 ml of distilled water.

4. Polish two copper electrodes and set up cell as shown in Figure 36.2.

5. Soak a $\frac{3}{8}$ in. $\times$ 3 in. (approximate dimensions) strip of filter paper in 0.5 M NaCl or KCl. When the salt bridge is inserted between the two solutions the voltmeter should register a voltage. Record this voltage initially and at $\sim$1 min intervals for several minutes.

Other qualitative observations that can be made:

1. If an ammeter is available that will measure 0–50 mA (= milliamps), measure the maximum current produced by the cell (i.e., replace the voltmeter by an ammeter). Try different widths of the filter paper salt bridges, changing the amount of electrode submerged, etc., and the concentration or type of the electrolyte solution used for the salt bridges, and observe any changes in the cell current.

2. Remove the voltmeter from the circuit and short the cell by connecting the anode and cathode leads. After a period of time ($\sim$30 min) observe the changes in appearance of the electrodes. Does either electrode show evidence of copper deposition?

---

*Of particular importance is the way in which the voltmeter is to be interpreted with respect to the *sign* of the cell emf.

**A number of complex ions can be studied using the same procedure given here, e.g., $Zn(NH_3)_4^{2+}$, $Cd(NH_3)_4^{2+}$. A particularly suitable group includes cyanide complexes, but the cyanide ion is not desirable for safety reasons.

III. DATA ANALYSIS

The cell is composed of two half-reactions:

$$Cu(s) + 4NH_3(5.4\ M) = Cu(NH_3)_4^{2+}(0.5\ M) + 2e^- \quad \text{(anode)}$$
$$\underline{Cu^{2+}(0.5\ M) + 2e^- = Cu(s) \hspace{5.5cm} \text{(cathode)}}$$
$$Cu^{2+}(0.5\ M) + 4NH_3(5.4\ M) = Cu(NH_3)_4^{2+}(0.5\ M)$$

From the Nernst equation

$$E_{cell} = E^\circ_{cell} - \frac{0.059}{2} \log \left[\frac{a_{Cu(NH_3)_4^{2+}}}{a_{Cu^{2+}}(a_{NH_3})^4} \right]. \tag{8}$$

If we assume that

$$a_{Cu(NH_3)_4^{2+}} \sim a_{Cu^{2+}}$$
$$a_{NH_3} \sim 5.4\ M$$

then E°_{cell} may be calculated using Equation 8 from the observed emf, E_{cell}. Using the relation between ΔG°_{cell} and E°_{cell},

$$\Delta G^\circ_{cell} = -E^\circ_{cell} nF \tag{9}$$

($F = 23{,}060$ cal/V-mol, Faraday's constant), calculate ΔG°_{cell}, which is ΔG° for the reaction

$$Cu^{2+}(aq) + 4NH_3(aq) = Cu(NH_3)_4^{2+}(aq)$$

(with all activities equal to unity). Also, using the thermodynamic relation

$$\log K_f = - \frac{\Delta G^\circ}{2.3RT}, \tag{10}$$

obtain your estimate of the formation constant for $Cu(NH_3)_4^{2+}$.

IV. ERROR ANALYSIS

1. Suppose the error in the measured voltages in this experiment is ± 0.05 V. What is the resulting uncertainty in K_f?

2. In evaluating Equation 8 it was assumed that the activity of Cu^{2+} was approximately the same as $Cu(NH_3)_4^{2+}$. Suppose in fact that $a_{Cu^{2+}} \simeq 0.1 \, a_{Cu(NH_3)_4^{2+}}$ (i.e., the activity coefficients for these two ions are quite different at $0.5 \, M$). What change in E°_{cell} and K_f would result from this correction?

3. The Nernst equation in the form of Equation 8 assumes the temperature to be 25°C. Calculate the change in E°_{cell} from a temperature variation of ± 5°C.

4. In the procedure it is emphasized that the electrode surfaces should be clean. Suppose that the copper electrode in the NH_3 solution has a coating of cuprous oxide (Cu_2O), such that the electrode reaction is

$$Cu_2O(s) + 4NH_3 = Cu(NH_3)_4^{2+} + CuO + 2e^-.$$

Qualitatively, how would the presence of this couple change the E°_{cell} from the value for two clean Cu electrodes? If thermodynamic data can be obtained for the process $2Cu + \frac{1}{2}O_2 = Cu_2O$, calculate the change in E°_{cell} caused by the oxide layer. (*Hint:* The change in the ΔG° for the net cell reaction will be required.)

5. *Qualitatively*, what will be the effect of a finite current on E_{cell}? As an extra experiment, the cell voltage can be measured across different resistances (starting with $> 1000\Omega$ to less than 10Ω), when the resistances are connected across the voltaic cell as follows:

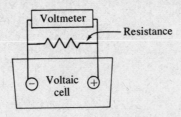

6. *Qualitatively*, what will the effect of the details of the salt bridge on the measured E_{cell} be? As an extra experiment, the cell voltage can be measured using different electrolytes in the salt bridge (e.g., compare $1.0\,M$ NaCl, KCl, NH_4Cl), different sizes of filter paper strips, or different concentrations of electrolytes (e.g., compare $0.5\,M$, $1.0\,M$, saturated NaCl).

NAME   **DATE**

SECTION

SELF-STUDY QUESTIONS

1. Describe what is meant by a redox reaction.

2. Write the Nernst equation and define all symbols used.

3. Given the half-reactions in Table 36.1, calculate E°_{cell} for a Fe/Ag couple producing Fe^{2+} ions and an Ag/Br_2 couple. Identify which electrode would be the anode. Sketch the cell construction for these couples (with the anode on the left).

4. Sketch the voltaic cell implied by the symbol

$$Ag|Ag^+(0.5M)\|Ag(NH_3)_2^+(1.0M),NH_3(1.0M)|Ag.$$

5. Calculate E°_{cell} and E_{cell} for the voltaic cell in Equation 4, assuming that the activities may be approximated by the concentrations and given that $K_f = 1.66 \times 10^7$.

6. What is the role of the salt bridge in a voltaic cell?

7. Why is it necessary for the voltmeter to draw very little current in order for the voltage measurements to have thermodynamic significance?

8. Since concentrated NH_3 (14.8 M) is used to make up the cell, why is the concentration of free NH_3 taken to be 5.4 in the data analysis?

APPENDIX 36. THE THERMODYNAMICS OF VOLTAIC CELLS

The change in free energy that accompanies any process provides a measure of the maximum amount of "useful work" that may be extracted from that process. In the case of an ideal voltaic cell the maximum electrical work that can be extracted from the cell reaction is

$$n\,FE_{\text{cell}} \tag{11}$$

where n = number of electrons transferred per mole of reaction
F = Faraday's constant, number of coulombs/mol (96,500)
E_{cell} = emf of cell.
By the usual convention ΔG, the change of free energy, is given by

$$\Delta G = -(\text{maximum useful work})$$

or

$$\Delta G_{\text{cell}} = -nFE_{\text{cell}}. \tag{12}$$

It is important to realize that Equation 10 is an idealization in the sense that for an operating cell the actual flow of current results in irreversible heating and concentration gradients, which increase entropy changes and diminish E_{cell}.* For this experiment, however, little current will be allowed to flow and it is reasonable to equate the observed cell emf to E_{cell} in Equation 10.

For a general thermodynamic process such as $a\text{A} + b\text{B} = y\text{Y} + z\text{Z}$, the change in free energy may also be written

$$\Delta G = \Delta G^{\circ} + 2.3\,RT\,\log\frac{\text{activity of products}}{\text{activity of reactants}} \tag{13}$$

$$= \Delta G^{\circ} + 2.3\,RT\,\log\frac{a_P}{a_R}$$

where

$$a_P = (a_{\text{Y}})^{y}(a_{\text{Z}})^{z} \tag{14}$$
$$a_R = (a_{\text{A}})^{a}(a_{\text{B}})^{b}.$$

The quantity a_{A} is the "activity" of the species A, and likewise for the quantities a_{B}, a_{Y}, a_{Z}. The activity is generally related to concentration, and in the limit of "ideal" solutions (which requires high dilution for electrolyte solutions), $a_{\text{A}} \longrightarrow [\text{A}]$. In the case of cell processes of the type discussed in this experiment, we may regard ΔG as the sum of the ΔG's for the two half-reactions, i.e.,

$$
\begin{array}{lll}
\text{A}^{n+} + ne^{-} = \text{A(s)} & \quad \Delta G_{\text{A}} & \tag{15}\\
\text{B(s)} = \text{B}^{n+} + ne^{-} & \quad -\Delta G_{\text{B}} & \\
\hline
\text{net reaction: A}^{n+} + \text{B(s)} = \text{A(s)} + \text{B}^{n+} & \quad \Delta G_{\text{net}} = \Delta G_{\text{A}} - \Delta G_{\text{B}} &
\end{array}
$$

---

*One extreme example of this kind of effect would be if the partition in Figure 36.1 allowed rapid mixing of the solutions, in which case a bulk reaction would occur on the electrode without any current passing between electrodes. The useful work obtained would be zero in this case.

Since ΔG_{net} depends on the sum of free energy changes, it is reasonable to write Equation 11 in the form

$$\Delta G_{net} = \Delta G_A^\circ + 2.3 \; RT \log \frac{a_{A(s)}}{a_{A^{n+}}} \tag{16}$$

$$- \left(\Delta G_B^\circ + 2.3 \; RT \log \frac{a_{B(s)}}{a_{B^{n+}}} \right).$$

ΔG_{net} is the free energy change available in the operation of the cell, and is equivalent to ΔG_{cell} in Equation 10, hence

$$\Delta G_{net} = -nFE_{cell} \tag{17}$$

$$= -nF \left[\left(E_A^\circ - 2.3 \frac{RT}{nF} \log \frac{a_{A(s)}}{a_{A^{n+}}} - E_B^\circ + 2.3 \frac{RT}{nF} \log \frac{a_{B(s)}}{a_{B^{n+}}} \right) \right]$$

Equation 14 is equivalent to the Nernst equation, Equation 5 of the Introduction. Thus the Nernst equation is a particular example of the use of Equation 11 to describe free energy changes.

EXPERIMENT

37

Voltaic Cells and the Nernst Equation: Determination of K_{sp} for $PbCl_2$

I. INTRODUCTION*

This experiment is very similar to Experiment 36 except that the equilibrium constant to be determined is to be the solubility product of $PbCl_2$. The basic scheme is as follows: the emf for the cell

$$Pb(s)|Pb^{2+}(1.0M)\|Cu^{2+}(1.0M)|Cu(s)$$

will be determined, and then NaCl or KCl will be added in excess to the Pb^{2+} solution, precipitating $PbCl_2$ and leaving an NaCl or KCl solution. The E_{cell} will be observed to change. The change in E_{cell} can be taken to reflect the change in the activity of Pb^{2+} from the original cell to the value

$$a_{Pb^{2+}} = \frac{K_{sp}^{PbCl_2}}{(a_{Cl^-})^2}. \tag{1}$$

Equation 1 is the standard solubility product expression. Note that the cell emf of the original cell is given by

$$E_{cell}^{original} = E_{Pb/Cu}^{\circ} - \frac{0.059}{2} \log \frac{a_{Pb^{2+}}}{a_{Cu^{2+}}} \simeq E_{Pb/Cu}^{\circ} \tag{2}$$

where we have assumed that $a_{Pb^{2+}}/a_{Cu^{2+}} \sim 1$ (i.e., at equal concentrations we assume the activities of Pb^{2+} and Cu^{2+} to be the same). $E_{Pb/Cu}^{\circ}$ is the standard emf for the $Pb|Pb^{2+}\|Cu^{2+}|Cu$ cell. For the KCl cell we find

$$E_{cell}^{KCl} = E_{Pb/Cu}^{\circ} - \frac{0.059}{2} \log \left[\frac{K_{sp}^{PbCl_2}}{a_{Cu^{2+}}(a_{Cl^-})^2} \right] \tag{3}$$

$$= E_{Pb/Cu}^{\circ} - \frac{0.059}{2} \log K_{sp}^{PbCl_2} + \frac{0.059}{2} \log (a_{Cu^{2+}}(a_{Cl^-})^2).$$

*The Introduction to Experiment 36 should be read.

In terms of measured quantities

$$\log K_{sp}^{PbCl_2} = \frac{-2(E_{cell}^{KCl} - E_{cell}^{original})}{0.059} + \log (a_{Cu^{2+}}(a_{Cl^-})^2) \tag{4}$$

from which $K_{sp}^{PbCl_2}$ can be determined. One problem with Equation 4 is that the contribution from the $\log (a_{Cu^{2+}}(a_{Cl^-})^2)$ term is difficult to estimate. In most cases the activity of an ion is less than its concentration, so that the upper limit of this term will be found by substituting the concentrations of each ion for the activities. The "lower limit" of this term is to use the approximate activity coefficients for the ion, i.e., $a_{Cu^{2+}} = \gamma_{Cu^{2+}}[Cu^{2+}]$.*

II. PROCEDURE

A cell like that in Figure 36.2 will be required for this experiment, as will solution volumes of 20–25 ml of 1.0 M Pb(NO$_3$)$_2$ and 1.0 M CuSO$_4$ for the two compartments. Record the volume of Pb(NO$_3$)$_2$ added to the beaker and leave enough volume in the beaker for the addition of solid NaCl or KCl later (5–10 ml). Clean the Pb and Cu electrodes with sandpaper (or the equivalent). Prepare ~10 ml of saturated NaCl or KCl solution** and soak a ⅜ in. × 3 in. strip of filter paper as the salt bridge.

1. Prepare the above cell, and measure the voltage. After a few minutes take another voltage measurement. A small amount of PbCl$_2$ precipitate will have formed from the salt bridge, but the change in Pb^{2+} concentration is too slight to be important.

2. From the volume of Pb(NO$_3$)$_2$ in the cell compute the amount of NaCl or KCl that must be added to precipitate all Pb^{2+} and to have a saturated† NaCl or KCl solution (solubility is approximately 36 g/100 ml of H$_2$O and 35 g/100 ml, respectively). Weigh out this amount of NaCl or KCl and add slowly to the Pb side of the cell, with stirring (gently). A precipitate of PbCl$_2$ will form, and a change in the voltage will be observed. Continue stirring until the voltage has become constant and it may be reasonably expected that as much salt has dissolved as can (3–4 min). Record the final voltage, plus observations of the way in which the voltage changed upon addition of the precipitation reagent.

Other qualitative observations that can be made:

1. Using an ammeter with a 0–50 mA sensitivity, measure the current of the cell before and after the addition of chloride. If the change of current is strictly ohmic, then the ratio of currents will be in the ratio of cell emfs. However, the change in

---

*The activity coefficients are themselves functions of concentration, and depend on the nature of the ion (especially the charge and coordination number of the ion). Generally there is significant uncertainty in the estimate of activities for the ionic species under the conditions of this experiment, and hence significant uncertainty will be present in the value of $K_{sp}^{PbCl_2}$ obtained.

**Other electrolytes can be used for the salt bridge, such as NH$_4$NO$_3$, that will not precipitate Pb^{2+}.

†It may be preferred to add enough chloride that the final chloride concentration is approximately 1 M, rather than a saturated solution. The activity of a 1 M NaCl or 1 M KCl solution is approximately 0.66 and 0.61, respectively.

the electrolyte concentration can change the efficiency of the salt bridge on the surface of the Pb electrode, such that the cell's "internal resistance" is changed.

2. The chloride can be added in measured amounts, such that not all Pb^{2+} is precipitated by the first few additions. If the cell emf is plotted versus moles of chloride added, the resulting graph will be equivalent to a "potentiometric titration." As a separate experiment (and a more convenient way to make this observation), a solution of chloride can be added to the Pb^{2+} solution using a buret and plotting emf versus volume of chloride solution added.

III. DATA ANALYSIS

The fundamental equation for the determination of $K_{sp}^{PbCl_2}$ is Equation 4, i.e.,

$$\log K_{sp}^{PbCl_2} = \frac{-2(E_{cell}^{KCl} - E_{cell}^{original})}{0.059} + \log (a_{Cu^{2+}}(a_{Cl^-})^2).$$

The two emfs are those directly measured. The activities must be estimated from the concentration. The approximate activity coefficients for Cu^{2+} and Cl^- in the concentration range used in this experiment are 0.07 and 0.6, respectively; hence, as a very rough estimate

$$a_{Cu^{2+}} = 0.07[Cu^{2+}] \tag{5}$$
$$a_{Cl^-} = 0.6[Cl^-].$$

Report the value of $K_{sp}^{PbCl_2}$ obtained by ignoring the activities and the value obtained by the estimate in Equation 5.

IV. ERROR ANALYSIS

1. Suppose the error in the difference of measured voltages is ± 0.05 V. What is the resulting uncertainty in $K_{sp}^{PbCl_2}$?

2. Suppose there is an uncertainty of ± 20 percent in the activity coefficients of Cu^{2+} or Cl^-. What is the resulting uncertainty in $K_{sp}^{PbCl_2}$ assuming that the full Equation 4 (with Equation 5) is used to obtain $K_{sp}^{PbCl_2}$ (i.e., activities are *not* ignored)?

3. Suppose the concentration of Cl^- is in error by ± 10 percent. What will be the effect on the calculated $K_{sp}^{PbCl_2}$ value (similar to 2 above)?

4. If a chloride containing salt bridge is used in this experiment a small amount of $PbCl_2$ will precipitate. If 0.01 ml of saturated NaCl ($\sim$0.36 g/ml) is transferred to 20 ml of 1.0 M $Pb(NO_3)_2$, what change in Pb^{2+} will occur, and what will the effect on $E_{cell}^{original}$ be?

NAME   DATE

SECTION

SELF-STUDY QUESTIONS

1. Write the Nernst equation and define all symbols used.

2. Given the half-reaction in Table 36.1, calculate $E°_{cell}$ for a Fe/Cu couple (Fe^{2+} produced) and a Zn/O$_2$ (acid medium) couple. Sketch the cell construction for these couples. Identify the anode and the cathode and sketch the direction of electron flow.

3. Sketch the voltaic cell implied by the symbol

$$Fe|Fe^{3+}(0.1M)\|Fe(OH)_3, OH^-(0.1M)|Fe.$$

4. Calculate E°_{cell} and E_{cell} for the voltaic cell in 3, assuming that the activities may be approximated by the concentrations and given that $K_{sp} = 1.66 \times 10^{-37}$.

5. What is the role of the salt bridge in a voltaic cell?

6. If a voltmeter draws significant current will the measured cell emf be less than or greater than the "ideal" (or thermodynamic) emf? Why?